T

**Nelson Advanced Modular Science**

# Principles of Physical and Organic Chemistry

## ALAN JARVIS

Nelson

**Thomas Nelson and Sons Ltd**
Nelson House  Mayfield Road
Walton-on-Thames  Surrey
KT12 5PL  UK

First published by Thomas Nelson and Sons Ltd 1997

I(T)P  Thomas Nelson is an International Thomson Publishing Company
I(T)P  is used under licence

ISBN 0-17- 448257-4
NPN 9 8 7 6 5 4 3 2

Publication team:
Acquisitions: Mary Ashby/Chris Coyer
Administration: Jenny Goode
Editorial Management: Simon Bell
Freelance Editorial: Tim Jackson/Mary Korndorffer
Marketing: Jane Lewis
Production: Liam Reardon
Design: Maria Pritchard
Typesetting and illustration: Hardlines

Printed in China

## Photography acknowledgements

The author and publishers are grateful to the following for permission to reproduce photographs:

Fig 1.1, page 1, British Steel; Fig 1.5b, page 7, Alan Thomas; Fig 1.6, page 8, Alan Thomas; Fig 2.1, page 20, Adam Hart-Davis/Science Photo Library; Fig 2.2, page 20, Fred George/Tony Stone Worldwide; Fig 2.4, page 21, Alan Thomas; Fig 2.7, page 22, Alan Thomas; Fig 4.1, page 48 Robert Frerck/Tony Stone Images; Fig 4.2, page 42, Andy Ross Photography; Fig 4.3, page 49, Andy Ross Photography; Fig 4.4, page 60, Alan Thomas; Fig 4.6, page 62, Stuart Boreham; Fig 6.1, page 73, Arnulf Husmo/Tony Stone Worldwide; Fig 6.2, page 73, Andy Ross Photography; Fig 6.3b, page 75, Alan Thomas; Fig 6.4, page 76, Spectrum Colour Library; Fig 6.5, page 78, Alan Thomas; Fig 6.6, page 82, Andy Ross Photography; Fig 6.7, page 82, Andy Ross Photography; Fig 6.8, page 82, Bert Blokhuis/Tony Stone Images; Fig 7.1, page 84, Alan Thomas; Fig 7.3, page 87, Alan Thomas; Fig 9.1, page 102, Ann Ronan Picture Library/Image Select; Fig 9.2, page 103, Alan Thomas; Fig 9.3, page 103, Alan Thomas.

# Contents

# Introduction

This textbook is one of a series of four produced in response to a demand from students and their teachers for resource material in support of the chemistry courses which lead to examinations set by the University of London Examination and Assessment Council in the new modular format. There has been a widespread development of modular courses at Advanced level, and ULEAC (now Edexcel London Examinations) took this step in September 1994. There is also an ever-present pressure on syllabus writers to introduce new material into syllabuses to ensure that they reflect adequately the role of chemistry in society today, yet the principal core concepts laid down by common agreement and the School Curriculum and Assessment Authority must retain their rightful place. Writers of the new syllabus and these texts have endeavoured to balance these conflicting demands.

There is a bewildering variety of chemistry texts discussing aspects of the subject at an appropriate level for the A-level student, and it is not the intention of this series to divert the attention of students from these. Indeed it is hoped that students will be excited by their study of chemistry and will want to pursue specialist avenues of interest, as countless others have done in years past. However, it is recognised that at certain times students seek a text which will encapsulate in a relatively small volume the outline of necessary study for each of the Edexcel London Examinations modules in chemistry.

These volumes are written by the examiners, all experienced teachers, specifically to prepare students for these examinations, and all the necessary basic material of the syllabus is covered. They further prompt and give pointers for further study for the interested student.

We hope that students will find these texts helpful and supportive of their studies at A-level and their preparation for examinations, and also stimulating to further reading in a wider context.

**Geoff Barraclough**
Chemistry Series Editor

## The author
**Alan Jarvis** was Head of Chemistry at Stoke-on-Trent Sixth Form College and was a Chief Examiner in Chemistry for Edexcel London Examinations. Alan Jarvis died in July 1997.

# The Periodic Table

**Key**

| Atomic number |
|---|
| **Symbol** |
| Name |
| Molar mass in g mol$^{-1}$ |

| Period | Group 1 | Group 2 | | | | | | | | | | | Group 3 | Group 4 | Group 5 | Group 6 | Group 7 | Group 0 |
|---|---|---|---|---|---|---|---|---|---|---|---|---|---|---|---|---|---|---|
| 1 | 1 H Hydrogen 1 | | | | | | | | | | | | | | | | | 2 He Helium 4 |
| 2 | 3 Li Lithium 7 | 4 Be Beryllium 9 | | | | | | | | | | | 5 B Boron 11 | 6 C Carbon 12 | 7 N Nitrogen 14 | 8 O Oxygen 16 | 9 F Fluorine 19 | 10 Ne Neon 20 |
| 3 | 11 Na Sodium 23 | 12 Mg Magnesium 24 | | | | | | | | | | | 13 Al Aluminium 27 | 14 Si Silicon 28 | 15 P Phosphorus 31 | 16 S Sulphur 32 | 17 Cl Chlorine 35.5 | 18 Ar Argon 40 |
| 4 | 19 K Potassium 39 | 20 Ca Calcium 40 | 21 Sc Scandium 45 | 22 Ti Titanium 48 | 23 V Vanadium 51 | 24 Cr Chromium 52 | 25 Mn Manganese 55 | 26 Fe Iron 56 | 27 Co Cobalt 59 | 28 Ni Nickel 59 | 29 Cu Copper 63.5 | 30 Zn Zinc 65.4 | 31 Ga Gallium 70.4 | 32 Ge Germanium 73 | 33 As Arsenic 75 | 34 Se Selenium 79 | 35 Br Bromine 80 | 36 Kr Krypton 84 |
| 5 | 37 Rb Rubidium 85 | 38 Sr Strontium 88 | 39 Y Yttrium 89 | 40 Zr Zirconium 91 | 41 Nb Niobium 93 | 42 Mo Molybdenum 96 | 43 Tc Technetium (99) | 44 Ru Ruthenium 101 | 45 Rh Rhodium 103 | 46 Pd Palladium 106 | 47 Ag Silver 108 | 48 Cd Cadmium 112 | 49 In Indium 115 | 50 Sn Tin 119 | 51 Sb Antimony 122 | 52 Te Tellurium 128 | 53 I Iodine 127 | 54 Xe Xenon 131 |
| 6 | 55 Cs Caesium 133 | 56 Ba Barium 137 | 57 ▲ La Lanthanum 139 | 72 Hf Hafnium 178 | 73 Ta Tantalum 181 | 74 W Tungsten 184 | 75 Re Rhenium 186 | 76 Os Osmium 190 | 77 Ir Iridium 192 | 78 Pt Platinum 195 | 79 Au Gold 197 | 80 Hg Mercury 201 | 81 Tl Thallium 204 | 82 Pb Lead 207 | 83 Bi Bismuth 209 | 84 Po Polonium (210) | 85 At Astatine (210) | 86 Rn Radon (222) |
| 7 | 87 Fr Francium (223) | 88 Ra Radium (226) | 89 ▲▲ Ac Actinium (227) | 104 Unq Unnil-quadium (261) | 105 Unp Unnil-pentium (262) | 106 Unh Unnil-hexium (263) | | | | | | | | | | | | |

▲ Lanthanide elements

| 58 Ce Cerium 140 | 59 Pr Praseodymium 141 | 60 Nd Neodymium 144 | 61 Pm Promethium (147) | 62 Sm Samarium 150 | 63 Eu Europium 152 | 64 Gd Gadolinium 157 | 65 Tb Terbium 159 | 66 Dy Dysprosium 163 | 67 Ho Holmium 165 | 68 Er Erbium 167 | 69 Tm Thulium 169 | 70 Yb Ytterbium 173 | 71 Lu Lutetium 175 |
|---|---|---|---|---|---|---|---|---|---|---|---|---|---|

▲▲ Actinide elements

| 90 Th Thorium 232 | 91 Pa Protactinium (231) | 92 U Uranium 238 | 93 Np Neptunium (237) | 94 Pu Plutonium (242) | 95 Am Americium (243) | 96 Cm Curium (247) | 97 Bk Berkelium (245) | 98 Cf Californium (251) | 99 Es Einsteinium (254) | 100 Fm Fermium (253) | 101 Md Mendelevium (256) | 102 No Nobelium (254) | 103 Lr Lawrencium (257) |
|---|---|---|---|---|---|---|---|---|---|---|---|---|---|

# Energy – The driving force of life

## Introduction

Life on Earth has always depended on the availability of energy, either directly or indirectly, from the Sun. Modern industrial societies, however, are even more dependent on the ready availability of energy in various forms. Fuels such as natural gas, oil and coal are burned in enormous quantities in our homes, our transport and in industry. These fuels are used directly as sources of heat for cooking, driving engines, etc. and are also converted into other forms of energy, such as electricity, in order to provide light and to drive other kinds of machinery. Within our own bodies, energy is obtained in a variety of ways. In muscles, for example, energy is obtained from the hydrolysis of large phosphate-containing molecules.

All these energy-yielding processes are chemical reactions and this is reason enough for us to want to enquire into the relationship between chemical reactions and energy. In addition, however, a study of energy changes can lead to a better understanding of many of the fundamental processes in chemistry.

It is the study of the relationship between chemical reactions and heat changes which is the subject of this chapter, although chemical reactions can produce energy in many different forms such as heat, light, electricity etc. The relationship between chemical reactions and heat changes is often referred to as **thermochemistry** which is itself part of a larger study known as **thermodynamics**.

## Signs, symbols and terminology

### Enthalpy change

Heat changes have different values depending on the conditions under which they are measured, in particular whether they are measured at constant volume or constant pressure. The latter is more appropriate at this level since most reactions will be carried out in open containers, that is effectively at constant pressure. The heat change measured at constant pressure is known as the enthalpy change and is given the symbol $\Delta H$.

### Standard conditions

Standard conditions refer to the internationally agreed conditions under which an enthalpy change should be measured if it is to be called a standard enthalpy change (represented by the symbol $\Delta H^{\ominus}$). These conditions are a temperature of 298 K and a pressure of 1 standard atmosphere ($101\,325\,\mathrm{N\,m^{-2}}$).

### Units

Heat is a form of energy and both are measured in the same units. The unit used is that derived from the basic SI units, i.e. the joule (J), although this is rather small for most chemical reactions and the multiple unit the kilojoule (kJ) is

*Fig 1.1 Heavy industry, such as the steel industry, uses huge amounts of energy*

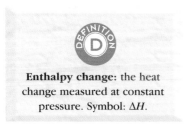

**Enthalpy change:** the heat change measured at constant pressure. Symbol: $\Delta H$.

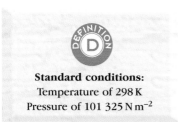

**Standard conditions:**
Temperature of 298 K
Pressure of $101\,325\,\mathrm{N\,m^{-2}}$

more frequently encountered. Another unit, known as the calorie, may be encountered occasionally, e.g. in food chemistry. This will not be used in this book but should it be met elsewhere, the conversion factor is 1 cal = 4.184 J.

## Sign convention

Chemical reactions which liberate heat energy to the surroundings are known as **exothermic** reactions. Such reactions are accompanied by an increase in the temperature of the surroundings. Chemical reactions which absorb heat energy from the surroundings are known as **endothermic** reactions. Such reactions are accompanied by a decrease in the temperature of the surroundings.

The sign convention used to distinguish between these two types of reaction is that a negative value for $\Delta H$ indicates an exothermic change, and a positive value for $\Delta H$ indicates an endothermic change.

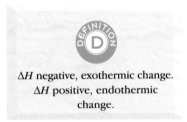

$\Delta H$ negative, exothermic change.
$\Delta H$ positive, endothermic change.

The sign is applied to the value of $\Delta H$ but does not indicate its magnitude. Thus a value of −400 kJ means 400 kJ of heat is evolved and is bigger than say −200 kJ which indicates that only 200 kJ of heat is evolved. Thus the enthalpy change for a reaction can be represented as follows:

$$C(s) + O_2(g) \rightarrow CO_2(g) \qquad \Delta H^{\ominus} = -393 \text{ kJ mol}^{-1}$$

This is interpreted to mean that when one mole of solid carbon is reacted with one mole of gaseous oxygen, gaseous carbon dioxide is formed and 393 kJ of heat are evolved when measured under standard conditions.

In writing equations such as this, it is essential that state symbols are shown since a change of physical state is accompanied by an enthalpy change. Indeed in some cases it is necessary to be even more specific about the actual substances used. For example the reaction above should more properly be written as:

$$C(\text{graphite, solid}) + O_2(g) \rightarrow CO_2(g) \qquad \Delta H^{\ominus} = -393 \text{ kJ mol}^{-1}$$

as the values for the other allotropes of carbon are different, e.g.

$$C(\text{diamond, solid}) + O_2(g) \rightarrow CO_2(g) \qquad \Delta H^{\ominus} = -395 \text{ kJ mol}^{-1}$$

A negative sign for $\Delta H$ thus indicates that energy has been *lost* from the substance involved and transferred to the surroundings.

## Enthalpy diagrams

Chemical reactions involve the breaking of some bonds and the making of others. The reagents have a certain amount of energy (or enthalpy) within them, as do the products. The enthalpy change of a reaction represents the difference in enthalpy between the reactants and products. This can be represented on an enthalpy diagram which has an increasing scale of energy as the vertical axis. This can be an actual scale with units specified or simply an arbitrary scale with unspecified units, e.g. for any reaction:

**Reactants (R) → Products (P)**

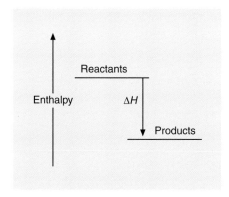

Fig. 1.2  Enthalpy diagram for an
exothermic reaction

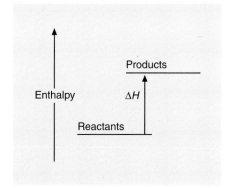

Fig. 1.3  Enthalpy diagram for an
endothermic reaction

the enthalpy diagram would be as shown in Figure 1.2 or Figure 1.3.

Since we have already designated the sign convention to be used, we have
defined the way in which $\Delta H$ is to be calculated.  Thus in all cases,

$$\Delta H = H_{Products} - H_{Reactants}$$

Thus the value of $\Delta H$ is negative for the exothermic reaction and positive for
the endothermic reaction, as required by the sign convention.

Another feature of enthalpy diagrams is that they clearly indicate whether the
reactants are more stable or less stable than the products.  Stability is related to
the position on the energy scale and the lower the position, the more stable
the substance is.  Thus in an exothermic reaction the products are more stable
than the reactants and the reactants are therefore said to be thermodynamically
less stable than the products.  The reverse is obviously true for endothermic
reactions.

## Changes of state

As stated previously, it is important to specify the physical state of the
substances involved when writing equations to represent an enthalpy change.
This is because any change in physical state is itself accompanied by an
enthalpy change.  The changes of state are named as in Figure 1.4.

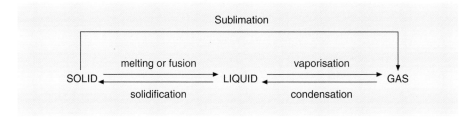

Fig. 1.4  Changes of state

Hence the **enthalpy of fusion** of  $H_2O$ would be the enthalpy change for the
reaction

$$H_2O(s) \rightarrow H_2O(l)$$

and the **enthalpy of vaporisation** would be the enthalpy change for the reaction

$$H_2O(l) \rightarrow H_2O(g)$$

Enthalpies of fusion, vaporisation and sublimation must always be endothermic since forces of attraction are being overcome. The reverse processes are exothermic.

## Enthalpies of combustion and formation

There are two particular types of reaction that are going to be more important to us in our studies than others. These are combustion reactions and reactions in which compounds are formed from their elements. The associated enthalpy changes are known as the **enthalpy of combustion** and the **enthalpy of formation** respectively.

These are defined as follows.

- The **standard enthalpy of combustion** $\Delta H_c^{\ominus}$ is the enthalpy change which occurs when one mole of a substance is completely burned in oxygen, under standard conditions.
- The **standard enthalpy of formation** $\Delta H_f^{\ominus}$ is the enthalpy change when one mole of a compound is formed from its elements in their standard states, under standard conditions.

The enthalpy of formation of any element in its standard state is zero.

Note that the combustion will not take place under standard conditions but the measurement of $\Delta H$ must be made when the conditions at the start and at the end of the reaction are standard.

### Enthalpy of combustion

The enthalpy of combustion is obviously associated with the process of burning or combustion in oxygen. The process is always exothermic. In writing equations, the product(s) of complete combustion need to be known, for example, when carbon is burned completely the product is carbon dioxide and not carbon monoxide, while for sulphur it is sulphur dioxide and not sulphur trioxide. Hence, the equation which represents the enthalpy of combustion of sulphur (rhombic) is:

$$S(\text{rhombic, solid}) + O_2(g) \rightarrow SO_2(g) \quad \Delta H^{\ominus} = -296.9 \, \text{kJ mol}^{-1}$$

The value for monoclinic sulphur would be different:

$$S(\text{monoclinic, solid}) + O_2(g) \rightarrow SO_2(g) \quad \Delta H^{\ominus} = -297.2 \, \text{kJ mol}^{-1}$$

Most organic compounds burn in oxygen, with hydrocarbons (which contain carbon and hydrogen only) and carbohydrates (which contain carbon, hydrogen and oxygen) being particularly important. The products of combustion for these compounds are always carbon dioxide and water. For example, methane (the principal constituent of natural gas), burns as follows:

**Standard enthalpy of combustion, $\Delta H_c^{\ominus}$**
The enthalpy change when one mole of a substance is completely burned in oxygen, under standard conditions.

**Standard enthalpy of formation, $\Delta H_f^{\ominus}$**
The enthalpy change when one mole of a compound is formed from its elements in their standard states, under standard conditions.

$$CH_4(g) + 2O_2(g) \rightarrow CO_2(g) + 2H_2O(g) \quad \Delta H^\ominus = -882 \, kJ \, mol^{-1}$$

Sucrose produces even more energy per mole:

$$C_{12}H_{22}O_{11}(s) + 12O_2(g) \rightarrow 12CO_2(g) + 11H_2O(g)$$
$$\Delta H^\ominus = -5644 \, kJ \, mol^{-1}$$

In reactions such as these, the water may be produced in the liquid or the gaseous state, depending on the temperature at which the reaction is carried out. If liquid water is produced, the values of the enthalpy changes will then be slightly different from those quoted above.

## Calorific values

Most fuels and foodstuffs now display a variety of information on the packaging, which includes the energy value of the fuel or food. These are based on enthalpies of combustion since combustion is the process that the fuel or food undergoes in the body. There is no reference to moles in the information given, since the fuel or food is not sold by moles but by weight or volume, nor is there any reference to a minus sign in the values although all are exothermic. British Gas, for example, quote a 'calorific' value of $38.4 \, MJ \, m^{-3}$ (megajoules per cubic metre; $1 \, MJ = 10^6 \, J$) and the cost to the consumer is calculated from the number of cubic metres used as measured by a gas meter installed on the premises.

Calorific values of foods are of vital importance for people who have to control their energy intake for dietary reasons. The values are displayed on the packaging as the number of units of energy which a given mass of food will provide on consumption. Some typical values are shown in Table 1.1, although actual values will depend on the exact brand used.

**Table 1.1** *The calorific value of some common foods, per 100g*

| Food | kJ | kcal |
|------|-----|------|
| Potatoes (boiled) | 342 | 82 |
| Potato crisps | 2350 | 562 |
| White bread | 1068 | 255 |
| Brown bread | 920 | 220 |
| Beef | 940 | 225 |
| Butter | 3031 | 724 |
| Sunflower oil spread | 2610 | 624 |

The energy value of foods eaten must equal the energy expenditure of our everyday activities. If we eat more than this, the body stores the excess in the form of fat, which is available for future use, but which is likely to cause numerous health problems. A 'calorie-controlled' diet involves eating foods with a lower energy value than the amount of energy expended. This forces the body to use up some of its reserves of fat in order to provide the additional energy required, and so weight is lost.

### *The relationship between enthalpy of formation and enthalpy of combustion.*

The enthalpy of formation refers to the formation of a compound from its elements, as defined earlier. When the compound formed is an oxide, this very often involves the same equation as the combustion of the element and consequently the enthalpy of formation of the oxide is the same as the enthalpy of combustion of the element.

For example, the enthalpy of combustion of graphite is the same as the enthalpy of formation of carbon dioxide ($-393\,\text{kJ mol}^{-1}$), since both are in fact the enthalpy change for the reaction

$$\text{C(graphite, solid)} + \text{O}_2(\text{g}) \rightarrow \text{CO}_2(\text{g}) \qquad \Delta H^\ominus = -393\,\text{kJ mol}^{-1}$$

This cannot be universally applied, however. For example, the equation

$$\text{C(diamond, solid)} + \text{O}_2(\text{g}) \rightarrow \text{CO}_2(\text{g}) \qquad \Delta H^\ominus = -395\,\text{kJ mol}^{-1}$$

represents the enthalpy of combustion of diamond but does not represent the enthalpy of formation of carbon dioxide since the carbon is not in its standard state. Graphite is the standard state for carbon since it is thermodynamically more stable than diamond at 298 K, as the enthalpy changes above show. Similarly the enthalpy of formation of lithium oxide is represented by the following equation

$$2\text{Li(s)} + \tfrac{1}{2}\text{O}_2(\text{g}) \rightarrow \text{Li}_2\text{O(s)} \qquad \Delta H^\ominus = -596\,\text{kJ mol}^{-1}$$

but this is not the enthalpy of combustion of lithium since two moles of lithium are involved in the equation.

## Experimental measurement of enthalpy changes

Although this topic is not specifically mentioned on the syllabus for module 2, some understanding of it is desirable and is sure to be encountered in school laboratories.

### Heat and temperature

The difference between these two terms must be made clear if the measurement techniques are to be understood. Temperature does not represent the amount of heat energy of a substance but merely the degree of hotness on some arbitrary scale. It is essentially a measure of the kinetic energies of the molecules or particles present in the substance and is independent of the amount of substance present. Heat, on the other hand, is a measure of the total energy in a substance and does depend on the amount of substance present. Thus comparing a large bucket of water at a temperature of 50 °C with say 250 cm$^3$ water in a beaker at 50 °C, both have the same temperature but the bucket of water would contain much more heat energy in total.

The amount of heat required to raise the temperature of 1 g of a substance by 1 K is called the **specific heat capacity** (*c*) and this differs from one substance

**Specific heat capacity is the amount of heat required to raise the temperature of 1 g of substance by 1 K.**

to another. Water, for example, has a value of about $4.2\,J\,g^{-1}\,K^{-1}$ while that for ethanol is about $2.4\,J\,g^{-1}\,K^{-1}$. The value for water is exceptionally high and hence water is able to store much more heat per unit mass than most other liquids.

Heat is always transferred from a hot substance to a cold substance and this will produce a change of temperature which can be measured. Thus, although there is no instrument which measures heat directly, the amount of heat transferred from a hot body to a cold body can be measured if the mass of the substance, its specific heat capacity and the temperature change are known. The heat transferred is then given by the expression:

**heat transfer = mass × specific heat capacity × temperature change**

**heat transfer $= m \times c \times \Delta T$**

Consistent units must of course be used and these will depend on the units given for $c$. These are usually $J\,g^{-1}\,K^{-1}$ in which case $m$ must be in grams. There is no need to change temperatures measured in °C into K since it is a temperature **difference** which is measured and a change of 1 °C is the same as a change of 1 K. The heat transfer will then be in joules. Measurement of heat therefore depends on transferring the heat to a known mass of another substance (usually water) and measuring the temperature rise. This is the basis of **calorimetry** and the apparatus used is called a **calorimeter**.

## Measurement of enthalpies of combustion

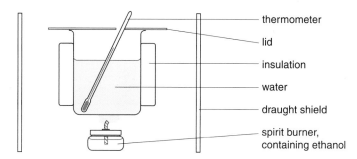

thermometer

lid

insulation

water

draught shield

spirit burner, containing ethanol

*Fig. 1.5a A simple form of calorimeter*

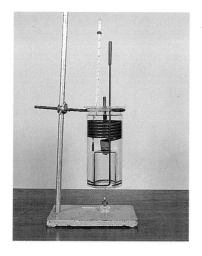

*Fig 1.5b A Thiemann calorimeter for measuring enthalpies of combustion*

A very simple form of calorimeter which could be used to measure the enthalpy of combustion of a liquid such as ethanol is shown in Figure 1.5. A known mass of water (150 g) is placed in a glass beaker or a metal can and its temperature noted (23 °C). A spirit burner containing ethanol is weighed and the burner is placed under the beaker and lit. The water in the beaker is stirred and after a certain time the final temperature of the water is noted (43 °C), the spirit burner is extinguished and it is re-weighed. The difference in mass of the spirit burner initially and at the end of the experiment gives the mass of ethanol burned (0.90 g). Thus:

$$\text{heat gained by water in calorimeter} = m \times c \times \Delta T$$
$$= 150 \times 4.2 \times 20$$
$$= 12600\,J$$

**Hence the heat produced by burning 0.90 g ethanol = 12.6 kJ**

# ENERGY – THE DRIVING FORCE OF LIFE

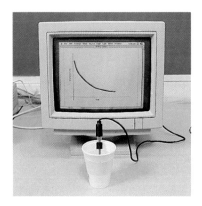

*Fig 1.6 Polystyrene cup calorimeter attached via datalogger to a computer, producing a graph which shows change in temperature over time*

Since 0.90 g is 0.90 /46 mol of ethanol, the heat produced by burning 1 mol of ethanol would be $12.6 \times 46/0.90 = 644$ kJ. Hence $\Delta H_c = -644$ kJ mol$^{-1}$. (The negative sign is inserted since the reaction must be exothermic, a temperature rise having occurred.)

This value is much less than the accepted value of $-1371$ kJ mol$^{-1}$. Some of the reasons for this are:

- Heat is still lost from the calorimeter to the surroundings despite the use of a lid and insulation of the sides.
- Some heat goes into the calorimeter instead of the water. This could be allowed for if the mass of the calorimeter and its specific heat capacity were known, or by calibrating the apparatus appropriately.
- Incomplete combustion of the ethanol due to an inadequate supply of oxygen, leading to the formation of products such as carbon monoxide or even carbon (as indicated by a deposit of soot on the bottom of the calorimeter).

More accurate methods are available for measuring enthalpies of combustion, e.g. using a bomb calorimeter, but these are beyond the scope of this module and will be found in textbooks of physical chemistry.

## Measuring enthalpy changes for reactions in solution

For reactions which take place in solution, the heat is generated (or absorbed) within the solutions themselves and it is simply a matter of keeping the heat within the solution. A calorimeter is therefore selected which is a very good insulator in order to reduce the heat escaping to the surroundings or being absorbed from the surroundings. A vacuum flask is very good in this respect, although in school laboratories a beaker made from expanded polystyrene is more frequently used. This material has a very low specific heat capacity and hence absorbs very little heat itself. The apparatus is shown in Figure 1.6 and good results can be obtained from this relatively simple apparatus. For example, the molar enthalpy change $\Delta H_N$ for the reaction:

$$HCl(aq) + NaOH(aq) \rightarrow NaCl(aq) + H_2O(l)$$

can be measured as follows.

Place 50 cm$^3$ of a 1.0 mol dm$^{-3}$ solution of hydrochloric acid in an insulated polystyrene cup and note its temperature. Add 50 cm$^3$ of 1.1 mol dm$^{-3}$ sodium hydroxide solution (an excess to ensure complete reaction) which is at the same temperature. Stir continuously and note the maximum temperature reached. The temperature rise will be about 6.5 °C. The total volume of solution will be 100 cm$^3$ which we will assume to be 100 g (although the actual mass could be found by weighing). Hence, assuming no heat losses to the surroundings and assuming that the specific heat capacity of the solution is the same as that of water (4.2 J K$^{-1}$ mol$^{-1}$), then:

$$\text{Heat absorbed by solution} = m \times c \times \Delta T$$

$$= 100 \times 4.2 \times 6.5$$

$$= 2730\,\text{J}$$

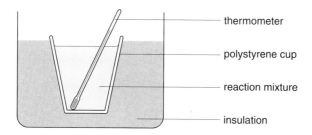

*Fig 1.7 Measuring enthalpy changes for reactions in solution*

This is the amount of heat that has been produced by reaction of 0.05 mol of acid. Hence 1 mol of acid would give 2730/0.05 J = 54600 J.

$$\therefore \Delta H_N = -54.6 \, \text{kJ mol}^{-1}$$

The negative sign is inserted since the reaction is obviously exothermic because a temperature rise was observed. The value obtained compares favourably with that obtained by more accurate methods of $-57.1 \, \text{kJ mol}^{-1}$.

## Temperature corrections

As seen above, reasonably accurate results can be obtained using the simple apparatus described. The results are less accurate, however, if the reaction being performed is slower than the neutralisation reaction used above. This is because there is still heat loss to the surroundings and this will increase if the reaction is slow because the heat will be lost over a longer period. This means that the temperature rise observed in the calorimeter is never a great as it should be. An allowance can be made for this by plotting a temperature–time graph. One reagent is placed in the polystyrene cup and its temperature noted at say 1 min intervals for say 4 min, stirring continuously. At a known time, say 4.5 min, the second reagent is added, stirring continuously, and the temperature noted more frequently until the maximum is reached. As the solution starts to cool the temperature is still recorded and stirring continued, for at least 5 min longer. A graph of temperature against time is then plotted. Graphs are given for an exothermic reaction (Figure 1.8) and for an endothermic reaction (Figure 1.9).

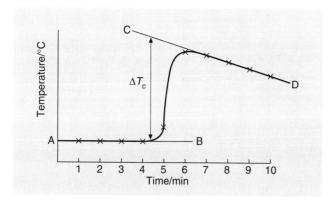

*Fig. 1.8 A temperature correction curve for an exothermic reaction*

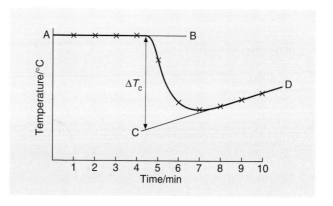

*Fig 1.9 A temperature correction curve for an endothermic reaction*

The corrected temperature rise or fall is then obtained from the graph by the following method:

- draw the best straight line through the results for the first 4 min (A–B)
- draw the best straight line through the results after the peak, when the temperature is falling (or rising) (C–D)
- continue both lines to the time of mixing (4.5 min in this case) and take the difference between them at this time.

This is the **corrected temperature rise** ($\Delta T_c$) and is greater than the maximum temperature observed for an exothermic reaction or the minimum observed for an endothermic reaction. The amount by which this exceeds the maximum (or minimum) observed is dependent on the steepness of the line after the maximum (or minimum) and this is dependent on the heat loss (or gain) from the surroundings. Thus the greater the heat losses to the surroundings, the more the temperature is corrected. Other methods of correcting temperature changes are available but this is the simplest and most commonly used at this level.

## Hess's law and the calculation of enthalpy changes

The enthalpy changes of many chemical reactions cannot be measured directly by the methods indicated in the previous section. In such cases it is possible to calculate the enthalpy change, given suitable alternative data. The calculations are based on an application of the First Law of Thermodynamics, one statement of which is: energy cannot be created or destroyed. It can only be converted from one form into another. Although this requires some modification in order to be absolutely true, it is a good working statement at this stage.

Two important deductions can be made from this. The first is that if the enthalpy change of a reaction is known, then the enthalpy change of the reverse reaction has the same value but with the sign changed. The second deduction is known as Hess's law, which states that the enthalpy change for a reaction is independent of the route by which the reaction is achieved, but depends only on the initial and final states.

If these statements were not true, it would be possible to create energy without any consumption of material. Desirable as this might be from the point of view of satisfying the worlds energy demands it is unfortunately impossible.

The first deduction is important since it allows us to know the enthalpy change for reactions which are very difficult or even impossible to measure. For example the enthalpy change for the reaction:

$$CO_2(g) \; \rightarrow \; C(\text{diamond, solid}) + O_2(g)$$

cannot be measured directly but its value must be $+395\,\text{kJ mol}^{-1}$ since the enthalpy change of the reverse reaction can be measured and has a value of $-395\,\text{kJ mol}^{-3}$.

Hess's law is important since it allows us to calculate enthalpy changes for reactions when these cannot be measured experimentally. Examples of how this can be achieved are given below.

**The First Law of Thermodynamics**
Energy cannot be created or destroyed. It can only be converted from one form to another.

**Hess's Law**
The enthalpy change for a reaction is independent of the route by which the reaction is achieved, but depends only on the initial and final states.

## Calculations using Hess's law

There are several different techniques for the performance of these calculations but all are essentially the same in that they require application of Hess's law.

### *Method 1: Constructing an alternative route*

Consider the conversion of reactants A into products B, a reaction for which it is impossible to measure $\Delta H$ directly. A can however be converted into B via two other compounds, C and D. This can be represented on the cycle shown in Figure 1.10. The calculation is then completed by application of Hess's law which states that the total enthalpy change for Route 1 = total enthalpy change for Route 2.

$$\therefore \Delta H = \Delta H_1 + \Delta H_2 + \Delta H_3$$

Hence $\Delta H$ can be calculated if values are known for $\Delta H_1$, $\Delta H_2$ and $\Delta H_3$.

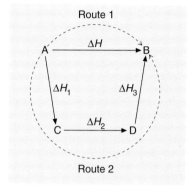

*Fig. 1.10 A reaction cycle*

### Example 1

Calculate the enthalpy change for the reaction C(s, graphite) → C(s, diamond), given the following data:

$$C(s, \text{ graphite}) + O_2(g) \rightarrow CO_2(g) \qquad \Delta H^\ominus = -393 \text{ kJ mol}^{-1}$$

$$C(s, \text{ diamond}) + O_2(g) \rightarrow CO_2(g) \qquad \Delta H^\ominus = -395 \text{ kJ mol}^{-1}$$

The alternative route for converting graphite to diamond is fairly obvious:

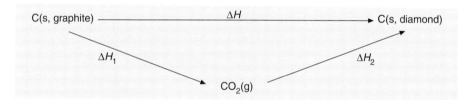

Application of Hess's law gives $\Delta H = \Delta H_1 + \Delta H_2$
From the data given $\Delta H_1 = -393 \text{ kJ mol}^{-1}$
$\Delta H_2 = +395 \text{ kJ mol}^{-1}$ (positive since the reaction in the data has been reversed).

Inserting appropriate values (including signs)

$$\Delta H = (-393) + (+395)$$

$$\Delta H = +2 \text{ kJ mol}^{-1}$$

Errors will be avoided if signs are only inserted with the values of $\Delta H$, rather than in the application of Hess's law.

Simple problems of this sort could also be solved by putting the data onto an enthalpy diagram. If this is done correctly, the answer becomes obvious from the diagram. Thus the data for the combustion of graphite and diamond can be put on to an enthalpy diagram as shown in Figure 1.9.

Graphite is therefore energetically more stable than diamond and it would take $+2\,kJ\,mol^{-1}$ to convert one mole of graphite to one mole of diamond. Similarly $2\,kJ\,mol^{-1}$ would be evolved if the reverse process was carried out.

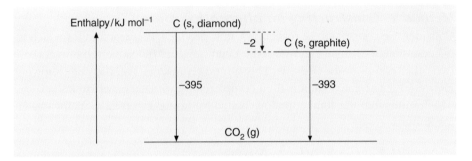

*Fig. 1.11  An enthalpy diagram for graphite and diamond*

In passing it might be noted that the conversion of graphite to diamond is not impossible, indeed many industrial diamonds are manufactured in this way. It is however an extremely difficult reaction to achieve, requiring very high temperatures and pressures. We might have expected such a difficult reaction to be endothermic but the value is so very small that this in itself is unlikely to make the reaction difficult to accomplish. There is also in fact a very high kinetic inhibition and this will be dealt with in Chapter 2.

### Method 2:  Combining equations

Considering the same example as above, the data given was:

$$C(s, graphite) + O_2(g) \rightarrow CO_2(g) \qquad \Delta H_1^{\ominus} = -393\,kJ\,mol^{-1} \qquad (1)$$

$$C(s, diamond) + O_2(g) \rightarrow CO_2(g) \qquad \Delta H_2^{\ominus} = -395\,kJ\,mol^{-1} \qquad (2)$$

Reversing equation (2) gives:

$$CO_2(g) \rightarrow C(s, diamond) + O_2(g) \qquad \Delta H_3^{\ominus} = +395\,kJ\,mol^{-1} \qquad (3)$$

Adding equations (1) and (3) gives:

$$C(s, graphite) + O_2(g) + CO_2(g) \rightarrow CO_2(g) + C(s, diamond) + O_2(g)$$

which on cancelling common species is the desired equation:

$$C(s, graphite) \rightarrow C(s, diamond)$$

and, doing the same to the enthalpy values as was done to the equations gives:

$$\Delta H = \Delta H_1^{\ominus} + \Delta H_3^{\ominus} = (-393) + (+395) = +2\,kJ\,mol^{-1}$$

Both methods are equally valid and the choice is up to the individual. However, the following points should be checked carefully whichever method is adopted.

- Check that the sign of the $\Delta H$ value is correct for the equation used.
- If more than one mole of a substance is involved, multiply the $\Delta H$ value accordingly.

**Example 2.**

Given the following data:

$$2NO_2(g) \rightarrow 2NO(g) + O_2(g) \qquad \Delta H^\ominus = +109\,kJ\,mol^{-1}$$

$$\tfrac{1}{2}N_2(g) + \tfrac{1}{2}O_2(g) \rightarrow NO(g) \qquad \Delta H^\ominus = +90.0\,kJ\,mol^{-1}$$

$$N_2(g) + 2O_2(g) \rightarrow N_2O_4(g) \qquad \Delta H^\ominus = +8.0\,kJ\,mol^{-1}$$

Calculate the enthalpy change for the reaction:

$$N_2O_4(g) \rightarrow 2NO_2(g)$$

*Using Method 1.*

The alternative cycle would be:

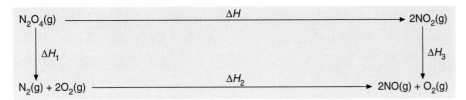

Applying Hess's law to this cycle gives:

$$\Delta H = \Delta H_1 + \Delta H_2 + \Delta H_3$$

$$\Delta H_1 = -(+8)$$
$$\Delta H_2 = 2(+90)$$
$$\Delta H_3 = -(+109)$$

Substituting values:

$$\Delta H = -(+8) + 2\,(+90) - (+109) = +63\,kJ\,mol^{-1}$$

*Using Method 2.*
*Data:*

$$\tfrac{1}{2}N_2(g) + \tfrac{1}{2}O_2(g) \rightarrow NO(g) \qquad \Delta H^\ominus = +90.0\,kJ\,mol^{-1} \qquad (4)$$

$$N_2(g) + 2O_2(g) \rightarrow N_2O_4(g) \qquad \Delta H^\ominus = +8.0\,kJ\,mol^{-1} \qquad (5)$$

$$2NO_2(g) \rightarrow 2NO(g) + O_2(g) \qquad \Delta H^\ominus = +109\,kJ\,mol^{-1} \qquad (6)$$

Reverse equation (5):

$$N_2O_4(g) \rightarrow N_2(g) + 2O_2(g) \qquad \Delta H^\ominus = -8.0\,kJ\,mol^{-1} \qquad (7)$$

Multiply equation (4) by 2:

$$N_2(g) + O_2(g) \rightarrow 2NO(g) \qquad \Delta H^\ominus = 2(+90.0)\,kJ\,mol^{-1} \qquad (8)$$

Adding equations (7) and (8) gives:

$$N_2O_4(g) \rightarrow 2NO(g) + O_2 \qquad\qquad\qquad\qquad (9)$$

Then for equation (9)

$$\Delta H^{\ominus} = 2(+90.0) + (-8) = +172\,\text{kJ mol}^{-1}$$

Reverse equation (6) and add to equation (9) give:

$$N_2O_4(g) \rightarrow 2NO_2(g) \tag{10}$$

Then for equation (10)

$$\Delta H^{\ominus} = (+172) + (-109) = +63\,\text{kJ mol}^{-1}$$

## Method 3: Using standard enthalpies of formation

This is really just a variation of method 1 but has the advantage that it can be used in all situations provided that values of enthalpies of formation are given. It is based on the idea that it must always be possible to construct an alternative route via the elements in their standard states since equations must balance in terms of the nature and number of atoms of each element involved. Thus any reaction may be written as:

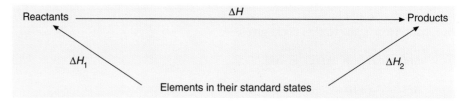

$\Delta H_2$ = the sum of the enthalpies of formation of the products

$\Delta H_1$ = the sum of the enthalpies of formation of the reactants

Applying Hess's law gives:

$$\Delta H = -\Delta H_1 + \Delta H_2$$

This leads to the universally applicable formula:

$$\Delta H = [\text{the sum of } \Delta H_f^{\ominus} \text{ (products)}] - [\text{the sum of } \Delta H_f^{\ominus} \text{ (reactants)}]$$

In applying this formula it must be remembered that:
- the enthalpies of formation of elements are arbitrarily taken to be zero;
- if more than one molecule of a substance is involved, the enthalpy of formation must be multiplied appropriately;
- the sign of the enthalpy change must be inserted with the value.

**Example 3.**
Calculate the standard enthalpy change for the combustion of ammonia in pure oxygen which occurs according to the equation:

$$4NH_3(g) + 3O_2(g) \rightarrow 2N_2(g) + 6H_2O(g)$$

given that the standard enthalpies of formation for $NH_3(g)$ and $H_2O(g)$ are $-46.1$ and $-242\,\text{kJ mol}^{-1}$, respectively.

Using the statement of Hess's law derived above:

$\Delta H^\ominus$ = the sum of $\Delta H^\ominus_f$ (products) – the sum of $\Delta H^\ominus_f$ (reactants)

$\therefore \Delta H^\ominus = \{6 \times (-242) + 2 \times 0\} - \{4 \times (-46) + 3 \times 0\}$

$\therefore \Delta H^\ominus = -1268 \, \text{kJ mol}^{-1}$

Note that the units are $\text{kJ mol}^{-1}$ despite the fact that there are 4 mol of ammonia in the equation and 6 mol of water. The units in this case refer to 1 mol of equation as written.

This problem could be equally well answered by application of methods 1 and 2.

## Average bond enthalpies

Covalent molecules can be broken up into the atoms of which they are composed by supplying energy in some form such as heat, light etc. and the enthalpy change for such a reaction can be measured. This is known as the **enthalpy of dissociation** which is defined as the enthalpy change when one mole of a substance is broken up into free gaseous atoms. For example the enthalpy of dissociation of hydrogen is the enthalpy change for the reaction:

$H_2(g) \rightarrow H(g) + H(g)$ $\qquad \Delta H^\ominus = +432 \, \text{kJ mol}^{-1}$

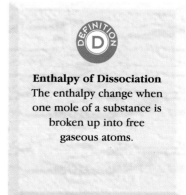

**Enthalpy of Dissociation**
The enthalpy change when one mole of a substance is broken up into free gaseous atoms.

Since the atoms formed are in the gaseous state then the covalent bonds holding the atoms together in the molecule can be considered to have been completely broken. Hence this enthalpy of dissociation is a measure of the strength of the covalent bonds in one mole of molecules of $H_2$ and is also referred to as the **bond enthalpy** of hydrogen, represented by $E(\text{H–H}) = +432 \, \text{kJ mol}^{-1}$. The values are always positive.

For polyatomic molecules such as methane ($CH_4$), the enthalpy of dissociation would be the enthalpy change for the reaction:

$CH_4(g) \rightarrow C(g) + 4H(g)$ $\quad \Delta H^\ominus = +1664 \, \text{kJ mol}^{-1}$

In this reaction, four covalent carbon–hydrogen bonds are broken and it would seem reasonable to assume therefore that the amount of energy required to break one such bond would be $1664/4 = 416 \, \text{kJ mol}^{-1}$. This is only an average or mean value however and is known as the **average bond enthalpy** represented by $E(\text{C–H}) = +416 \, \text{kJ mol}^{-1}$.

The specific bond enthalpies for the four C–H bonds are in fact quite different from one another, e.g.

$CH_4(g) \rightarrow CH_3(g) + H(g)$ $\quad \Delta H^\ominus = +427 \, \text{kJ mol}^{-1}$

$CH_3(g) \rightarrow CH_2(g) + H(g)$ $\quad \Delta H^\ominus = +371 \, \text{kJ mol}^{-1}$

The reasons for this are beyond the scope of this book and so we need only concern ourselves with average bond enthalpies.

## Average bond enthalpies and enthalpy of reaction

Average bond enthalpies can be used to calculate the enthalpy change for a reaction. This is done by assuming that an alternative route for all reactions can be achieved theoretically via the gaseous atoms of the elements involved in the compounds. It is therefore a specific application of Hess's law to the cycle:

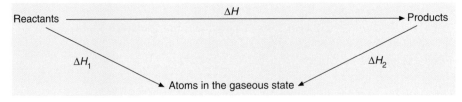

$\Delta H_1$ = sum of the average bond enthalpies of the reactants

$\Delta H_2$ = sum of the average bond enthalpies of the products

Applying Hess's Law gives:

$$\Delta H = \Delta H_1 - \Delta H_2$$

This leads to the universally applicable formula:

$\Delta H$ = [the sum of the average bond enthalpies of the reactants]

– [the sum of the average bond enthalpies of the products]

The method does require some knowledge of the bonding present in the particular molecules.

### Example

Calculate the enthalpy change for the reaction:

$$CH_4(g) + Cl_2(g) \rightarrow CH_3Cl(g) + HCl(g)$$

given the average bond enthalpies in $kJ\,mol^{-1}$:

$E(C–H) = 412; E(C–Cl) = 338; E(Cl–Cl) = 242; E(H–Cl) = 431$

$$\therefore \Delta H^\ominus = [4 \times E(C–H) + E(Cl–Cl)] - [3 \times E(C–H) + E(C–Cl) + E(H–Cl)]$$
$$= [(4 \times 412) + 242] - [(3 \times 412) + 338 + 431]$$
$$= [1890] - [2005] = -115\,kJ\,mol^{-1}$$

### Alternative method

This involves a simple accounting procedure:

Energy absorbed (bonds broken) = $(4 \times 412) + 242 = +1890\,kJ$

Energy evolved (bonds formed) = $(3 \times 412) + 338 + 431 = -2005\,kJ$

Total enthalpy change = energy absorbed + energy evolved
$$= 1890 - 2005 = -115\,kJ\,mol^{-1}$$

The arithmetic can be simplified if it is realised that three of the C–H bonds broken in $CH_4$ are re-formed in $CH_3Cl$. Hence in total only one C–H bond is broken together with one Cl–Cl bond and one C–Cl bond is formed together with one H–Cl bond. Hence:

$$\Delta H^\ominus = [E(\text{C–H}) + E(\text{Cl–Cl})] - [E(\text{C–Cl}) + E(\text{H–Cl})]$$
$$= [412 + 242] - [338 + 431]$$
$$= 654 - 769 = -115\,\text{kJ mol}^{-1}$$

# Questions

1   (a) Define   (i) enthalpy of formation;
                       (ii) enthalpy of combustion.

   (b) When 12.00 g of each of carbon and hydrogen are completely burned in oxygen, 393.5 and 1715.4 kJ are evolved respectively. Calculate the enthalpies of combustion of carbon and hydrogen.

2   (a) State Hess's law and give two examples to illustrate its usefulness in chemistry. No numerical data are required.

   (b) Calculate the enthalpy change for the following reaction, using the data below:

$$P_4O_{10}(s) + 6H_2O(l) \rightarrow 4\,H_3PO_4(s)$$

   *Data:* The enthalpies of formation of $P_4O_{10}(s)$, $H_2O(l)$ and $H_3PO_4(s)$ are –2984, –285.9 and –1279 kJ mol$^{-1}$, respectively.

3   (a) Construct a suitable cycle and apply Hess's law to calculate the enthalpy of formation of butane ($C_4H_{10}$) using the following data:

   Enthalpy of combustion of graphite  = –393.5 kJ mol$^{-1}$

   Enthalpy of combustion of hydrogen  = –285.9 kJ mol$^{-1}$

   Enthalpy of combustion of butane  = –2877.1 kJ mol$^{-1}$

   (Clue: write an equation for each of the enthalpies of combustion in the data and for the formation of butane. Then construct an alternative route for the formation of butane using the equations for the combustions.)

4   (a) How is the enthalpy of formation of a substance connected to the stability of the substance?

   (b) Sulphur has two allotropes, rhombic and monoclinic, both of which form sulphur dioxide on burning. The enthalpy of combustion of rhombic sulphur is –296.6 kJ mol$^{-1}$ and that of monoclinic sulphur is –297 kJ mol$^{-1}$. Suggest why these two values differ and deduce which of the allotropes is the more stable thermodynamically. Show the results of your deduction on an enthalpy diagram.

   Calculate the enthalpy change for the reaction:

$$\text{S(rhombic)} \rightarrow \text{S(monoclinic)}$$

**5**  Data:

| Substance | $H_2O(l)$ | $CO_2(g)$ | ethane, $C_2H_6(g)$ | ethene, $C_2H_4(g)$ |
|---|---|---|---|---|
| $\Delta H_f^{\ominus}$/kJ mol$^{-1}$ | −285.5 | −393 | −83.6 | +52.0 |

(a)  Write equations for the complete combustion of:

   (i)  ethane;

   (ii)  ethene;

   (iii)  hydrogen.

(b)  Calculate the enthalpy of combustion in each case.

(c)  From the results obtained in (b), calculate the enthalpy change for the reaction:

$$C_2H_4(g) + H_2(g) \rightarrow C_2H_6(g)$$

**6**  Data:     $4NH_3(g) + 5O_2(g) \rightarrow 4NO(g) + 6H_2O(l)$     $\Delta H = -1170\,kJ$

$4NH_3(g) + 3O_2(g) \rightarrow 2N_2(g) + 6H_2O(l)$     $\Delta H = -1530\,kJ$

$2H_2(g) + O_2(g) \rightarrow 2H_2O(l)$     $\Delta H = -576\,kJ$

(a)  Calculate the enthalpy of formation of ammonia.

(b)  Calculate the enthalpy of formation of nitrogen monoxide, NO(g).

**7**  Data:     $CH_4(g) + 2O_2(g) \rightarrow CO_2(g) + 2H_2O(l)$     $\Delta H = -890\,kJ\,mol^{-1}$.

$2CO(g) + O_2(g) \rightarrow 2CO_2(g)$     $\Delta H = -568\,kJ\,mol^{-1}$

$C(graphite) + O_2(g) \rightarrow CO_2(g)$     $\Delta H = -393\,kJ\,mol^{-1}$

$H_2(g) + \frac{1}{2}O_2(g) \rightarrow H_2O(l)$     $\Delta H = -285.5\,kJ\,mol^{-1}$

Use the data, as appropriate, to calculate:

(a)  the enthalpy of formation of methane;

(b)  the enthalpy of formation of carbon monoxide;

(c)  the enthalpy change when 1 mole of methane burns in a limited supply of oxygen to produce carbon monoxide and water.

**8** Data:

| Substance | $B_2H_6(g)$ | $B_2O_3(s)$ | $C_6H_6(g)$ | $CO_2(g)$ | $H_2O(g)$ |
|---|---|---|---|---|---|
| $\Delta H_f^\ominus$/kJ mol$^{-1}$ | +31.4 | −1270 | +83.9 | −393 | −242 |

Gaseous diborane, $B_2H_6$, and gaseous benzene, $C_6H_6$, combust in oxygen as follows:

$$B_2H_6(g) + 3O_2(g) \rightarrow B_2O_3(s) + 3H_2O(g)$$

$$C_6H_6(g) + 7.5O_2(g) \rightarrow 6CO_2(g) + 3H_2O(g)$$

Calculate which will produce the greater amount of heat, 50 kg of $B_2H_6(g)$ or 100 kg of $C_6H_6(g)$.

**9** Calculate the average bond enthalpy of a C-Cl bond given the following data:

$$C(graphite) \rightarrow C(g) \quad \Delta H = +715 \text{ kJ mol}^{-1}$$

$$Cl_2(g) \rightarrow 2Cl(g) \quad \Delta H = +242.2 \text{ kJ mol}^{-1}$$

$$C(graphite) + 2Cl_2(g) \rightarrow CCl_4(l) \quad \Delta H = -135.5 \text{ kJ mol}^{-1}$$

**10** Data:

| Bond | C–H | C–Br | Br–Br | H–Br |
|---|---|---|---|---|
| Average bond enthalpy/kJ mol$^{-1}$ | 413 | 209 | 193 | 366 |

Use the data to calculate the enthalpy change for the reaction:

$$CH_4(g) + Br_2(g) \rightarrow CH_3Br(g) + HBr(g)$$

**11** The enthalpy of formation of ammonia gas is −46 kJ mol$^{-1}$ and the average bond enthalpies of hydrogen gas and nitrogen gas are 436 and 945 kJ mol$^{-1}$, respectively. Comment on the relative values of these average bond enthalpies and attempt to explain the difference.

Use the data to calculate the average bond enthalpy of the N–H bond.

# Kinetics – how fast do reactions go?

A few simple test tube reactions demonstrate that reactions proceed at very different rates. Some, like the reaction between hydrochloric acid and sodium hydroxide solutions, are extremely fast. Others, like the reaction between a lump of calcium carbonate and dilute hydrochloric acid, will bubble away merrily for some time before the production of the bubbles of carbon dioxide gas ceases. Many reactions between organic compounds are very slow indeed and require heating for a considerable time. Some examples of such reactions will be found in Chapters 5 to 8. It is the object of this chapter to consider some of the factors which affect the rate of a chemical reaction and to consider how useful a study of the kinetics of a reaction can be to our understanding of the way in which chemical reactions occur.

Fig 2.1 Erosion of a stone lion outside Leeds town hall. An example of a slow reaction, in which acid rain, formed by sulphur dioxide and nitrogen dioxide dissolving in rainwater, reacts with the calcium carbonate of which limestone is largely composed, converting it into more soluble calcium sulphate, which is then gradually washed away

## What is meant by the rate of a reaction?

The rate of a reaction is measured by the change of reagent concentration in unit time. The concentration of reagent is measured in $mol\,dm^{-3}$ and hence the rate of reaction has units such as $mol\,dm^{-3}\,s^{-1}$, although other units of time may be more appropriate in some reactions.

## How is the rate of a reaction measured?

There are many different ways in which the rate of a reaction can be measured, some of which are outlined below.

### Withdrawal of samples and titration

In order to conform explicitly with the definition of rate, the amount of reagent which has been used up in a given time has to be measured. For example, in the reaction between a carboxylic acid and an alcohol to form an ester, such as:

$$CH_3CO_2H + CH_3CH_2OH \rightarrow CH_3CO_2CH_2CH_3 + H_2O$$

the initial concentration of the ethanoic acid $[A]_0$ is determined by titration with standard sodium hydroxide, using a suitable indicator such as phenolphthalein. The reagents are then heated under reflux at a constant temperature. After a known time a small sample is withdrawn with a pipette and immediately added to a known excess of standard sodium hydroxide solution, which prevents any further reaction. Titration with, for example, standard hydrochloric acid will allow the concentration of ethanoic acid remaining at time $t$, $[A]_t$, to be found. The concentration of ethanoic acid used up by that time is given by $[A]_0 - [A]_t$ and hence the average rate of reaction over the time of the experiment would then be $[A]_0 - [A]_t / t$. If, however, the reaction is allowed to continue and samples are withdrawn at a number of different time intervals, a number of different values for $[A]_t$ will be found. Plotting these as a function of time would give a graph as shown in Figure 2.3. The graph is a curve and the rate of reaction can be found from the gradient of the curve at any time $t$. Hence the rate of reaction is not constant but varies as

Fig 2.2 A much quicker reaction – exploding fireworks

the reaction proceeds since the concentrations of the reagents are constantly decreasing. The average rate over a specific time period can be measured (though this is not really of any significance) or the specific rate at a particular time can be measured. This latter is more useful as will be seen later and the most useful of all is the rate of reaction at zero time, also known as the **initial rate** of reaction, since this is the time when the concentrations of reagents are most accurately known.

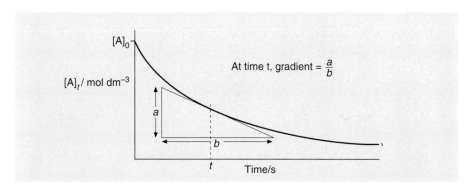

Fig. 2.3  Plot of $[A]_t$ against time. The slope of the curve at any time, t, gives the rate at that time

## Reactions producing gases

Reactions producing gases are most conveniently followed by measuring the volume of gas produced. Even though the gas is a product and not a reactant, its concentration (or volume) will increase at the same rate as the concentrations of the reactants decrease. For example, hydrogen peroxide decomposes in the presence of a catalyst, manganese(IV) oxide, according to the equation:

$$2H_2O_2(aq) \rightarrow 2H_2O(l) + O_2(g)$$

A sample of the hydrogen peroxide is placed in a conical flask in a thermostatically controlled water bath, and connected to a gas syringe as shown in Figure 2.4. At a known time, the catalyst is added and the volume of oxygen gas in the syringe noted at various time intervals. These volumes are plotted as a function of time, giving a graph as shown in Figure 2.6. The initial rate of the reaction is then the gradient of this graph at zero time.

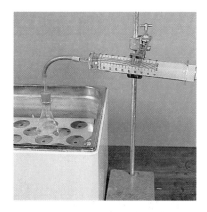

Fig 2.4 Syringe apparatus for measuring rate of reaction when one of the products is gaseous (see diagram in fig 2.5)

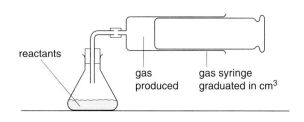

Fig. 2.5  A typical apparatus for measuring volumes of gases produced in a reaction

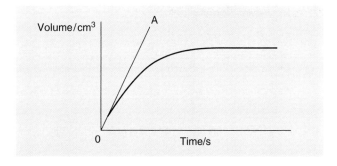

Fig. 2.6  Plot of volume of gas evolved against time. The gradient at zero time, OA, is the initial rate

## Reactions producing a colour change

If one of the reactants or one of the products of a reaction is coloured, the intensity of the colour can be used to measure the rate of reaction. For example, propanone ($CH_3COCH_3$) and iodine react together in the presence of an acid catalyst, to give iodopropanone and hydrogen iodide. The equation for the reaction is:

$$CH_3COCH_3(aq) + I_2(aq) \rightarrow CH_2ICOCH_3(aq) + HI(aq)$$

Iodine (a brown solution) is the only coloured substance in this reaction. Hence the rate of decrease of the colour of the solution is the same as the rate of decrease of the concentration of iodine and can be used to measure the rate of reaction. The colour intensity can be measured by an instrument known as a colorimeter which records the colour intensity as a reading on a meter or chart recorder. Alternatively, the colorimeter could be interfaced to a datalogger or microcomputer which would allow a graph of concentration of $I_2$ against time to be plotted, as shown in Figure 2.7, and this would be particularly useful for quicker reactions. The instrument will require calibration in order to establish the relationship between the reading on the meter and the concentration of the species being observed. The rate of this particular reaction could also be followed by the sampling and titration technique described above. The rate of this reaction is such that sampling at five minute intervals is appropriate. The samples are run rapidly into a flask containing sodium hydrogencarbonate which neutralises the acid catalyst and hence 'quenches' or 'freezes' the reaction. The iodine concentration can then be found at leisure by titration with standard sodium thiosulphate solution using starch as indicator. The colorimeter approach is more satisfactory since no sampling is necessary and the reading can be taken almost instantaneously if a datalogger or microcomputer is used.

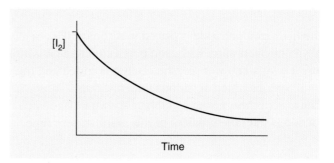

*Fig. 2.7 (a) Colorimeter interfaced with computer producing graph of colour intensity against time (b) Results from a datalogger*

## Using time as a measure of rate

In some reactions it is convenient to measure the time it takes for a particular stage of a reaction to be reached. For example, in the reaction between sodium thiosulphate solution and dilute hydrochloric acid, a precipitate of sulphur is produced:

$$S_2O_3^{2-}(aq) + 2H^+(aq) \rightarrow H_2O(l) + SO_2(g) + S(s)$$

The rate of this reaction can be measured as:

$$\text{rate} = \frac{\text{amount of sulphur formed}}{\text{time } (t)}$$

If the amount of sulphur formed in several experiments is the same then the top line is constant and:

$$\text{rate} \propto \frac{1}{t}$$

Also, if the amount of sulphur is small, then the rate will approximate to the initial rate. Hence the initial rates for this reaction, using different concentrations of, say, sodium thiosulphate, can be compared by comparing $1/t$ for each reaction. This experiment is often performed by placing the beaker containing the reaction mixture over a mark drawn on a piece of paper. The time is taken for sufficient sulphur to be produced to make the mark invisible when viewed from above the solution. Provided that the same total volume of solutions are mixed in each experiment, then it will take a given amount of sulphur to make the solution opaque each time and the time for this to occur in each case is inversely proportional to the rate of the reaction.

## 'Clock' reactions

Clock reactions are special reactions in which a coloured species is produced but the colour is not seen initially because another reagent is present which reacts with the coloured substance as fast as it is formed. When all this reagent has been used however then the colour will appear quite suddenly and with quite startling effect. Such a reaction is:

$$H_2O_2(aq) + 2H^+(aq) + 2I^-(aq) \rightarrow 2H_2O(l) + I_2(aq)$$

The colour of the iodine solution is brown but a better effect is seen if starch is added to the solution so that the colour produced is blue. On its own this reaction would produce a blue colour quickly and the intensity of this colour would then increase with time. If, however, a solution of sodium thiosulphate is added to the original reagents, the iodine formed will react with the thiosulphate ions in preference to the starch, forming a colourless solution:

$$I_2(aq) + 2S_2O_3^{2-}(aq) \rightarrow S_4O_6^{2-}(aq) + 2I^-(aq)$$

When all the thiosulphate ions have been used up, an immediate blue colour will appear as the iodine is now able to react with the starch.

Provided that the amount of thiosulphate is the same each time, the time for the appearance of this blue colour will be inversely proportional to the rate of the reaction, and will thus allow the rates of reaction for different concentrations of, say, hydrogen ions, to be compared.

Although several different methods of measuring rates of reaction have been considered, the list is by no means exhaustive and there are many other methods which might be used for measuring the rate of a reaction, such as measuring changes in pH or conductivity, etc.

## Factors which affect the rates of chemical reactions

There are six factors which can affect the rate of a chemical reaction:
- **concentration** of reactants in solution
- **pressure** of any gases present
- **surface area** of any solid reactants
- **temperature**
- **catalysts**
- **light**

An increase in pressure, temperature or surface area of a solid will lead to an increase in the rate of reaction. This is also true of the concentration of most reactants – though not all, as will be seen later.

Catalysts are substances which are capable of increasing the rate of a reaction without being chemically changed themselves and will be dealt with later in this chapter.

Light only affects the rates of certain reactions, e.g. the reaction of chlorine with alkanes (see Chapter 6). Certain modern adhesives are based on this principle and do not begin to react until exposed to sunlight.

## Theories of reaction rates

Any theory which purports to explain how quickly reactions occur must be able to explain how the factors mentioned above affect the rate of reaction in the way they do. There are two main theories of kinetics, the **collision theory** and the **transition state** theory.

### The collision theory

The collision theory is based on the simple concept that before two particles can react they must collide. Only a small fraction of the total number of collisions, however, results in a reaction. There are two main reasons for this:
- the molecules must approach each other in the correct orientation (this is sometimes called the **steric factor**)
- the molecules must have a certain minimum amount of energy (often referred to as the **activation energy**).

Reaction rates can therefore be increased if either collisions occur more frequently and/or the proportion of molecules having the required activation energy can be increased.

Thus if the concentration or pressure is increased then there are more particles in a given volume and they are bound to collide more frequently. Hence the increase in rate is explained simply by the increase in the number of collisions. Similarly, an increase in the surface area of a solid would lead to more collisions between the solid surface and the other reactant and hence the rate would increase.

In the case of an increase in temperature, the explanation is a little more complex since the kinetic energies of the molecules are increased along with the frequency of collisions. The former is by far the greater effect, however, and results in more molecules having energies greater than the minimum energy required. These more energetic molecules are bound to collide more frequently even if there is no increase in the overall frequency of collision, and hence the rate of reaction increases.

The molecules in a gas (or a liquid) do not all have the same kinetic energies since they do not all have the same speeds. The way in which their energies are distributed is shown in the so called **Maxwell–Boltzmann distribution**. Figure 2.8 shows this distribution for a given sample of gas at three different temperatures $T_1$, $T_2$ and $T_3$ where $T_3 > T_2 > T_1$. From this it can be seen that at any temperature, only a few molecules have very low or very high energies, most being around the most common value, represented by the peak value. As the temperature increases, the curve broadens out, the peak value decreasing and moving towards a higher energy value. The area under the curve represents the total number of molecules in the sample and is therefore constant. The area under the curve beyond the activation energy $E_{Act}$ represents the number of molecules having energies greater than or equal to the activation energy. This increases as temperature increases and so therefore does the rate of reaction.

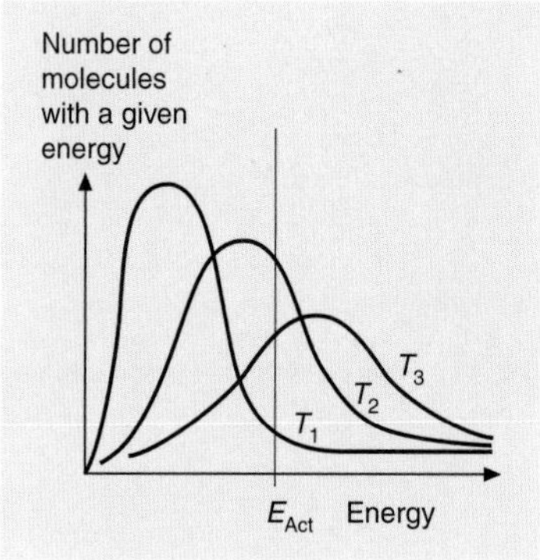

*Fig. 2.8 The Maxwell–Boltzmann distributions for a sample of gas at three different temperatures*

## The transition state theory

This theory considers the details of the actual collision between two molecules. As two molecules approach each other repulsion between their electron clouds will push them apart again unless they have sufficient kinetic energy to overcome this repulsion. If they do get sufficiently close to each other, a rearrangement of electron clouds will occur so that some bonds are broken and new bonds form. While this is happening, a highly unstable species is formed for a very short period of time in which some bonds are partially broken and others partially formed. This unstable species is known as the **transition state** or **activated complex**. During this process, the kinetic energy of the collision is converted into potential energy which can be shown on an enthalpy diagram usually referred to as the **reaction profile** (Figures 2.9 and 2.10). The activated complex occurs at the peak of this profile and the energy gap between the reactants and this peak is known as the activation energy $E_{Act}$ for the reaction. Molecules must have this activation energy, or greater, when they collide if they are to react successfully, since this amount of energy must be absorbed even though energy is given out on going from transition state to products. Thus the activation energy can be considered as a barrier to reaction and the greater its value the slower the reaction will be. An example of a reaction involving a transition state is the $S_N2$ reaction of a primary haloalkane with aqueous alkali. This reaction is discussed in further detail in Chapter 7.

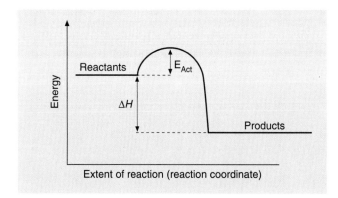

*Fig. 2.9 The reaction profile for an exothermic reaction*

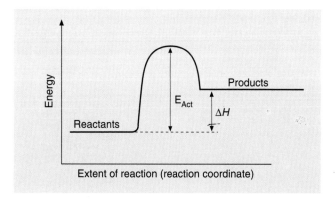

*Fig. 2.10 The reaction profile for an endothermic reaction*

## Thermodynamic and kinetic stability

Reference was made in Chapter 1 to thermodynamic stability, which is determined by the enthalpy levels of the reactants and products of a reaction. In an exothermic reaction, the products are considered to be more stable than the reactants and the reactants are therefore classified as being thermodynamically less stable. We would normally expect that such a reaction would be spontaneous, i.e. likely to occur in the forward direction. This is in fact not always true and other factors (such as entropy) need to be taken into account in determining whether or not a reaction is spontaneous. We shall, however, adopt the simplification that a negative $\Delta H$ is one of the principal driving forces of a reaction and so indicates the direction in which the reaction is likely to take place.

However, such reactions frequently do not occur in practice and this is often due to the fact that the activation energy for the reaction is too high under the conditions being used. Reactions with a high activation energy are said to be kinetically stable. An example of such a reaction is

$$C(s, \text{graphite}) + O_2(g) \rightarrow CO_2(g) \qquad \Delta H = -393 \, \text{kJ mol}^{-1}$$

Graphite is therefore thermodynamically unstable and the reaction would be expected to be spontaneous from left to right. In practice, no reaction occurs at room temperature because the reactants are kinetically stable. At an elevated temperature the reaction proceeds as expected since the reactants now have sufficient energy to overcome the activation energy.

Conversely, there is nothing to prevent certain endothermic reactions occurring provided that there is sufficient energy available to overcome both the enthalpy change required and the activation energy. Reactions also appear to go in the direction in which there is an increase in randomness or degree of disorder, but a study of this is beyond the scope of this volume.

## Catalysts

**Catalysts** are substances which increase the rate of a chemical reaction whist remaining chemically unchanged themselves.

Catalysts are substances which increase the rate of a chemical reaction whilst remaining chemically unchanged themselves. They usually work by providing an alternative route for the reaction to occur which has a lower activation energy than the normal route. Thus more molecules have enough energy to overcome the activation energy for the alternative route and hence the reaction proceeds faster. The effect of a catalyst on the reaction profile of a reaction is shown in Figure 2.11. Note that the introduction of a catalyst has no effect on the enthalpy change of the reaction.

Frequently the catalyst may allow the reaction to proceed in two or more steps. For example, the reaction between peroxodisulphate(VI) ions and iodide ions:

$$S_2O_8{}^{2-}(aq) + 2I^-(aq) \rightarrow 2SO_4{}^{2-}(aq) + I_2(aq)$$

is catalysed by $Fe^{2+}$ ions. The catalyst is believed to function by allowing two steps to occur thus:

$$S_2O_8{}^{2-}(aq) + 2Fe^{2+}(aq) \rightarrow 2SO_4{}^{2-}(aq) + 2Fe^{3+}(aq)$$

followed by:

$$2Fe^{3+}(aq) + 2I^-(aq) \rightarrow 2Fe^{2+}(aq) + I_2(aq)$$

Note that the catalyst ($Fe^{2+}$ ions) is regenerated in the reaction and is not therefore used up. A possible reaction profile for this process is shown in Figure 2.12.

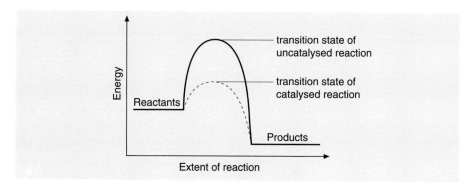

*Fig. 2.11 The effect of a catalyst on the profile of a reaction*

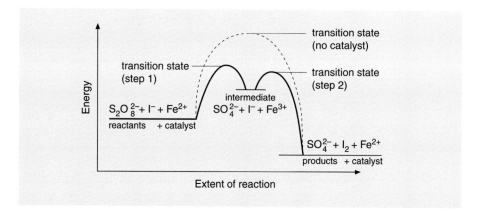

*Fig. 2.12 A possible reaction profile for a reaction catalysed by $Fe^{2+}$ ions*

# The rate expression and order of reaction

The **rate expression** is a more precise mathematical way of expressing how the concentrations of reactants affect the rate of reaction. The way in which the concentration of a reactant affects the rate of a chemical reaction can only be determined by practical investigation. As a result of such experiments a relationship between the rate of reaction and the concentrations of the various reagents, known as the rate expression, can be determined.

For a reaction such as:

$$M + N \rightarrow Y$$

The rate expression is:

$$rate = k[M]^a[N]^b$$

$a$ = the order of the reaction with respect to reactant M,
$b$ = the order of the reaction with respect to reactant N.
The **overall order** of the reaction is the sum of the individual orders, $a + b$ in this case. $k$ is called the **rate constant**.

For example if $a = 1$ and $b = 2$ in the above expression, then the reaction is said to be first order with respect to M, second order with respect to N and third order overall. Hence for the reaction:

$$BrO_3^-(aq) + 5Br^-(aq) + 6H^+(aq) \rightarrow 3Br_2(aq) + 3H_2O(l)$$

experiment shows that the reaction is:
- first order with respect to $BrO_3^-(aq)$
- first order with respect to $Br^-(aq)$
- second order with respect to $H^+(aq)$
- fourth order overall.

Note that there is *no* relationship between the numbers of moles of a reagent in the overall chemical equation and the powers to which the concentrations are raised in the rate expression.

The order of reaction does not have to be a whole number; it can be fractional and it can also be zero.

If a reaction is first order with respect to a given reagent, doubling the concentration of that reagent would double the rate of reaction.

For a reaction which is second order with respect to a given reagent, doubling the concentration of that reagent would multiply the rate of reaction by a factor of four.

For a reaction which is zero order with respect to a given reagent, doubling the concentration of that reagent would have no effect on the rate of reaction.

Thus changing the concentration of a reagent does not always increase the rate of reaction and changing some concentrations can be more effective than others.

## The rate constant

The rate constant is a factor in the rate expression which is temperature dependent. It is also related to the activation energy of the reaction, which is also temperature dependent. The rate constant and the activation energy are inversely related (by the Arrhenius equation), so that a high activation energy means that the rate constant is small and vice versa.

Units for the rate constant vary according to the precise rate expression in which it occurs. For a reaction which is first-order overall, e.g.

$$\text{rate} = k[A] \text{ and hence } k = \frac{\text{Rate}}{[A]}$$

The units for rate will be $mol\,dm^{-3}\,s^{-1}$ and for [A] $mol\,dm^{-3}$. Hence the units of $k$ will be $\dfrac{mol\,dm^{-3}\,s^{-1}}{mol\,dm^{-3}}$ , which ends up as $s^{-1}$.

For an overall second order reaction, however, e.g.

$$\text{rate} = k[C][D] \text{ and hence } k = \frac{\text{Rate}}{[C][D]}$$

Hence the units of $k$ are $\dfrac{\text{mol dm}^{-3}\,\text{s}^{-1}}{(\text{mol dm}^{-3})^2}$, which ends up as $\text{mol}^{-1}\,\text{dm}^3\,\text{s}^{-1}$.

## Experimental determination of order

Determination of the order of a reaction is achieved by measuring the rate of reaction for varying concentrations of one reagent while keeping the concentrations of other reagents constant as well as maintaining a constant temperature.

Concentrations of other reagents are maintained at levels which are more or less constant (i.e. change by only a very small amount) by using a large excess of these reagents. In other words, this is achieved by using concentrations which are very large compared to the concentration of the reactant under investigation.

For example, if the reaction:

**A + B → products**

is under investigation to find the order of reaction with respect to reactant A and the rate is being measured for a concentration of A of $0.01\,\text{mol dm}^{-3}$, then a concentration of B of say $1.00\,\text{mol dm}^{-3}$ might be used. Thus at the end of the experiment the concentration of A will be 0 but the concentration of B will still be $1.00 - 0.01 = 0.99\,\text{mol dm}^{-3}$, which is virtually constant.

The results of such experiments can be interpreted in one of several different ways.

## *Method 1. Using graphs of concentration against time*

Graphs of concentration against time are obtained from experiments in which the concentrations of a reactant A are found at various time intervals, as described earlier in this chapter. The shapes of such graphs are shown in Figure 2.13 for zero-order reactions, Figure 2.14 for first-order reactions and Figure 2.15 for second order reactions. Only if the reaction is zero order with respect to A will the graph of concentration versus time be linear, that is, the slope or gradient of the line is constant at all times. If this is the case, no further interpretation is necessary. If it is not so then one of the other methods will be needed.

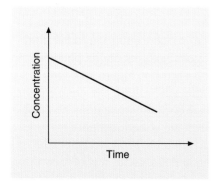

*Fig. 2.13 A concentration/time graph for a zero-order reaction*

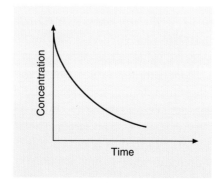

*Fig. 2.14 A concentration/time graph for a first-order reaction*

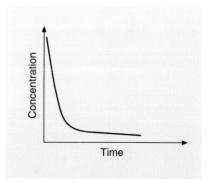

*Fig. 2.15 A concentration/time graph for a second-order reaction*

### *Method 2. Using graphs of rate against concentration.*

If the graph in method 1 is not linear then the gradient can be taken at various time points to find the rate of reaction at these times. A graph of rate of reaction against concentration can then be plotted. Only if the reaction is first order will the graph of rate versus [A] be linear since Rate $\alpha$ [A] (Figure 2.17).

If the reaction is not first order then a curve will result. In this case it may be worth plotting a graph of rate against $[A]^2$, which will be linear if the reaction is second order (Figure 2.18).

As an alternative, if the order is not 1, 2 or 0, a graph of log(rate) against log[A] will always give a straight line, the gradient of which gives the order of the reaction and the intercept on the *y*-axis is log (rate constant). This is true because, in general:

$$\text{Rate} = k[A]^n$$

Hence:

$$\log \text{Rate} = n \log[A] + \log k$$

Which is of the form:

$$y = mx + c$$

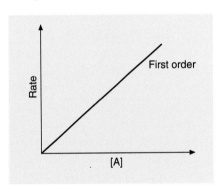

Fig. 2.16 A graph of rate against concentration, linear for a first-order reaction

Fig. 2.17 A graph of rate against square of concentration, linear for a second-order reaction

### *Method 3: Using successive half-lives*

The **half-life** of a reaction is the time taken for the concentration to fall from any selected value to half of that value. Successive half-lives can be obtained from a concentration versus time graph. This is shown in Figure 2.18 for a first order reaction. In the first order reaction, all the half-lives are the same and are independent of concentration. In the case of second order reactions, the half lives become successively larger. The idea of half-life is most commonly encountered in studies of radioactive decay, which always follows first-order kinetics. It can apply to any reaction, however, and is a useful way of identifying first order reactions.

The half-life of a reaction is the time taken for the concentration to fall from any selected value to half of that value.

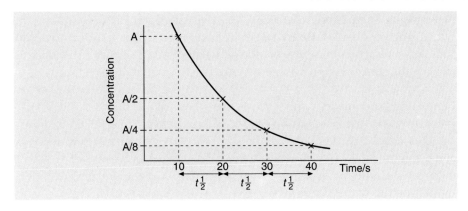

*Fig. 2.18 Successive half-lives for a first-order reaction*

## Method 4. The initial rate method

The initial rate method involves a series of experiments in which the initial rate of reaction is measured for different known concentrations of one reagent, say A. This could be done by determining the concentrations of A at known times and plotting a graph of concentration against time, as before. The tangent to this graph is found at zero time and this gives the initial rate of the reaction. The experiment is repeated using a different concentration of the reagent 'A' but keeping the temperature and the concentrations of other reagents and catalyst constant. The same procedure can then be used for the other reagents in turn. A comparison of the initial rates with the initial concentrations of each reagent will then allow the order to be found. Other ways of measuring the initial rate can be used, as is shown in the example below for the reaction:

$$2H_2(g) + 2NO(g) \rightarrow 2H_2O(g) + N_2(g)$$

This reaction lends itself to a novel way of measuring the rate experimentally which we have not seen before. Since four moles of gaseous reactants gives three moles of gaseous products, the rate could be measured by following the decrease in pressure at constant temperature. From this it is possible to calculate the rate of nitrogen production in $mol\,dm^{-3}\,s^{-1}$. The initial rates for a set of six experiments using different initial concentrations of reactants at 1073 K are shown in Table 2.1.

**Table 2.1**

| Experiment | Initial concentration/mol dm$^{-3}$ | | Initial rate of N$_2$ production/ mol dm$^{-3}$ s$^{-1}$ |
|---|---|---|---|
| | [NO] | [H$_2$] | |
| 1 | $6.00 \times 10^{-3}$ | $1.00 \times 10^{-3}$ | $3.19 \times 10^{-3}$ |
| 2 | $6.00 \times 10^{-3}$ | $2.00 \times 10^{-3}$ | $6.36 \times 10^{-3}$ |
| 3 | $6.00 \times 10^{-3}$ | $3.00 \times 10^{-3}$ | $9.56 \times 10^{-3}$ |
| 4 | $1.00 \times 10^{-3}$ | $6.00 \times 10^{-3}$ | $0.48 \times 10^{-3}$ |
| 5 | $2.00 \times 10^{-3}$ | $6.00 \times 10^{-3}$ | $1.92 \times 10^{-3}$ |
| 6 | $3.00 \times 10^{-3}$ | $6.00 \times 10^{-3}$ | $4.30 \times 10^{-3}$ |

Experiments 1 and 2 show that when $[H_2]$ is doubled and $[NO]$ is constant, the rate of reaction doubles. Hence rate $\alpha$ $[H_2]$ and the reaction must be first order with respect to $H_2$. A comparison of Experiments 3 and 1 confirms this since trebling the concentration trebles the rate.

Comparing Experiments 4 and 5 shows that when $[NO]$ is doubled while $[H_2]$ is constant, the rate increases by a factor of four. Hence rate $\alpha$ $[NO]^2$ and the reaction must be second order with respect to NO. Comparing Experiments 4 and 6 confirms this since multiplying the concentration by a factor of 3 increases the rate by a factor of 9, that is $3^2$. Hence the reaction is third order overall and the rate expression is:

$$\text{rate} = k[NO]^2[H_2]$$

Substituting any set of values will then give the value of the rate constant $k$. Thus using the figures from Experiment 1:

$$k = \frac{\text{rate}}{[NO]^2[H_2]}$$

$$= \frac{3.19 \times 10^{-3}}{(6.00 \times 10^{-3})^2 \, (1.00 \times 10^{-3})}$$

$$= \frac{3.19 \times 10^{-3}}{3.6 \times 10^{-8}} = 8.86 \times 10^4 \, mol^{-2} \, dm^6 \, s^{-1}$$

As can be seen this procedure is relatively easy if the order is a simple integer. Other cases require a more sophisticated mathematical approach which is beyond the scope of A level.

## The stepwise nature of reactions

It has been stated in the previous section that there is seldom any link between the number of moles of a given reagent in the stoichiometric equation and the order with respect to this reagent. Where the numbers happen to be the same it is entirely coincidental. This may appear quite surprising at first until it is realised that all reactions, except a small number of very simple ones, proceed via a number of simple steps. Each step involves only one or two particles, and the stoichiometric equation is the summation of these steps. Each step proceeds at a given rate under the conditions specified but the reaction as a whole can only proceed at the rate at which the slowest step proceeds, this step being known as the **rate-determining step**.

It is therefore not quite so surprising that the order of the reaction is in fact linked not to the stoichiometric equation but to that step in the process which determines the overall rate. At A level it is usually assumed that the order of the reaction with respect to each reagent, from the rate expression, is the number of moles of each reagent in the rate-determining step, although this is an over-simplification as the following example shows.

Thus a knowledge of the orders of the various reagents in a reaction, together with a knowledge of the stoichiometric equation, can sometimes allow us to suggest what the various steps in a reaction are. For example, for the reaction:

$$BrO_3^- + 5Br^- + 6H^+ \rightarrow 3Br_2 + 3H_2O$$

Experiment shows that the rate equation is:

$$\text{rate} = k[BrO_3^-]^1[Br^-]^1[H^+]^2$$

This is fourth order overall but it is almost impossible statistically for there to be a simultaneous collision involving four particles ($BrO_3^- + Br^- + 2H^+$) as a rate-determining step. Also, collisions between particles of like charge are very unlikely.

We can conjecture what the likely steps are and in this case they are thought to be:

$$H^+ + Br^- \rightleftharpoons HBr \qquad \text{Fast}$$

$$H^+ + BrO_3^- \rightleftharpoons HBrO_3 \qquad \text{Fast}$$

$$HBr + HBrO_3 \rightarrow HOBr + HOOBr \qquad \text{Slow}$$

$$HOOBr + HBr \rightarrow 2HOBr \qquad \text{Fast}$$

$$HOBr + HBr \rightarrow H_2O + Br_2 \qquad \text{Fast}$$

Addition of the equations up to and including the slow (rate-determining) step, shows the number of reactant particles corresponding to the order for each reactant. This knowledge of the steps by which a reaction proceeds is usually called the **mechanism** of the reaction, particularly in organic chemistry. An example of the way in which this can be applied is given in Chapter 7.

## Questions

1   The diagram shows the reaction profile for a given reaction.

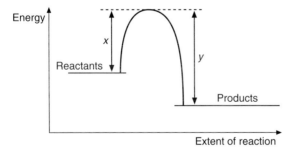

(a) State the following in terms of $x$ and $y$:

(i)  the activation energy of the forward reaction;

(ii)  the enthalpy change, $\Delta H$, of the forward reaction.

(b) Make a copy of the diagram and sketch on it a possible energy profile for the same reaction carried out in the presence of a catalyst.

(c) State the enthalpy change, $\Delta H_1$, of the catalysed forward reaction in terms of $x$ and $y$.

(d) Explain, with reference to the diagram, how the catalyst affects the rates of both the forward and the back reactions.

2 The graph below represents the distribution of molecular energies in a gas at a constant temperature $T$.

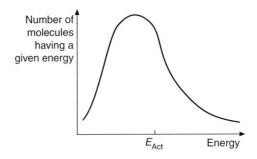

(a) Make a copy of the graph and sketch on it two curves, labelling them $T_1$ and $T_2$, showing the distribution of energies at temperature $T_1$ which is appreciably lower than $T$, and at temperature $T_2$ which is appreciably higher than $T$.

(b) $E_{Act}$ represents the activation energy of a chemical reaction at temperature $T$. Explain what is meant by the term 'activation energy'. Would its position change for the two curves which you have added to the graph? Explain your answer.

(c) Explain carefully the relevance these graphs have to explaining the effect of an increase in temperature on the rate of a chemical reaction.

3 The stoichiometric equation for the hydrolysis of thioethanamide in alkaline solution is:

$$CH_3CSNH_2 + 2OH^- \rightarrow CH_3CO_2^- + HS^- + NH_3$$

The rate of this reaction is found to be first order with respect to each of the reagents $CH_3CSNH_2$ and $OH^-$.

(a) What is meant by the term 'order' of reaction?

(b) Write the rate expression for this reaction.

(c) Deduce and explain the effect that a doubling of the concentration of hydroxide ions would have on the rate of reaction.

(d) Give a reason why the numbers of moles in the equation for the reaction are not the same as the orders for each reagent.

4  The following data refer to experiments conducted on the reaction of magnesium with dilute hydrochloric acid:

| Concentration HCl/mol dm$^{-3}$ | 0.3 | 0.4 | 0.6 | 0.8 | 1.2 |
|---|---|---|---|---|---|
| Time for formation of 20 cm$^3$ hydrogen/s | 240 | 136 | 60 | 34 | 15 |

In each case, the reaction was carried out at constant temperature of 20°C and 1 atm pressure. The same length of magnesium ribbon was used for each experiment and there was a large excess of hydrochloric acid.

(a)  Sketch the apparatus you would use to do these experiments.

(b)  Explain how the rate of this reaction is related to the times recorded in the table.

(c)  Why was a large excess of hydrochloric acid used in the experiments?

(d)  Why was a constant length of magnesium ribbon used in each experiment?

(e)  What is the order of reaction with respect to hydrochloric acid? Write an expression relating the rate of reaction to the concentration of HCl.

(f)  Sketch and label the reaction profile for the reaction between magnesium and hydrochloric acid.


5  In acid solution, bromate(V) ions slowly oxidise bromide ions to bromine:

$$BrO_3^- + 5Br^- + 6H^+ \rightarrow 3Br_2 + 3H_2O$$

The following experimental data was obtained in investigating the rate of this reaction at constant temperature. The solutions of bromate(V), bromide and hydrochloric acid used were all 1 mol dm$^{-3}$.

| Mixture | Volume of BrO$_3^-$/cm$^3$ | Volume of Br$^-$/cm$^3$ | Volume of HCl/cm$^3$ | Volume of water/cm$^3$ | Relative rate of formation of Br$_2$ |
|---|---|---|---|---|---|
| A | 50 | 250 | 300 | 400 | 1 |
| B | 50 | 250 | 600 | 100 | 4 |
| C | 100 | 250 | 600 | 50 | 8 |
| D | 50 | 125 | 600 | 225 | 2 |

(a)  Explain why a certain volume of water is added in each experiment.

(b)  Deduce the order of reaction with respect to each of the ions BrO$_3^-$, Br$^-$ and H$^+$. Explain your reasoning.

(c)  Write the rate expression for the reaction and use it to find the units of the rate constant.

# Equilibria – how far do reactions go?

## Reversible reactions

In studying chemical reactions so far, particularly in carrying out stoichiometric calculations, it has been assumed that the reactions go to completion. By this we mean that the reactants which are not in stoichiometric excess are totally used up and converted into products. For example, in calculating the mass of calcium oxide obtainable from heating 1 g of calcium carbonate at 900 °C, it must be assumed that the entire 1 g decomposes according to the equation:

$$CaCO_3(s) \rightarrow CaO(s) + CO_2(g)$$

and hence arrive at the answer of 0.56 g CaO.

If this reaction was performed in an open container this would in fact be the case. If, however, it was performed in a closed container, only a small proportion of the calcium carbonate would decompose, no matter how long the temperature was maintained at 900 °C. At this stage, the vessel would contain a certain amount of each of the three substances, which would not change provided that the temperature remained constant.

The reason for this is that the reaction is actually **reversible**, that is, the calcium oxide and the carbon dioxide also react together at this temperature to form calcium carbonate. Thus there are two reactions going on in opposite directions; obviously if the carbon dioxide were allowed to escape, the reverse reaction could not occur.  It is standard procedure in chemistry to regard the substances on the left hand side of an equation to be the reactants and those on the right to be the products. In this situation however, all substances present are really both reactants and products. Nevertheless we shall continue to refer to the substances on the left-hand side of the equation as the reactants. The reaction from left to right will be referred to as the forward reaction and the one from right to left as the reverse reaction.

## The nature of a chemical equilibrium

In any reversible reaction of this kind (assuming that we start with the substances on the left-hand side at a given temperature) the rate of the reaction from left to right will start at a certain level but will decrease as the concentrations of the reactants decrease. The reverse reaction will not start until some products have been formed, and even then the rate will be slow since their concentrations will be small. However, the rate of the reverse reaction will increase as the concentrations of the substances on the right-hand side of the equation increase (see Figure 3.1). Thus a point will be reached when the forward and the reverse reactions will be occurring at the same rate. At this point, the concentrations of the substances will remain constant but the reactions have not stopped. The reaction is said to have reached a state of **equilibrium** but it is a **dynamic equilibrium,** represented by the symbol $\rightleftharpoons$.

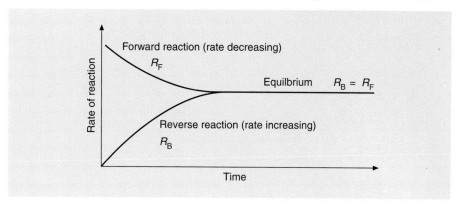

Fig. 3.1 *Changes in the rates of forward and reverse reactions approaching equilibrium*

A **dynamic equilibrium** is achieved when the forward reaction and the reverse reaction are occuring at equal rates.

There are two types of equilibria: **homogeneous**, where all the components are in the same physical state, and **heterogeneous**, where the components are in different physical states.

The reaction above is an example of a heterogeneous system and should be written with the appropriate equilibrium sign:

$$CaCO_3(s) \rightleftharpoons CaO(s) + CO_2(g)$$

Some examples of homogeneous systems are:

$$CH_3CO_2H(l) + C_2H_5OH(l) \rightleftharpoons CH_3CO_2C_2H_5(l) + H_2O(l)$$

$$2SO_2(g) + O_2(g) \rightleftharpoons 2SO_3(g)$$

$$H_2(g) + I_2(g) \rightleftharpoons 2HI(g)$$

It should be noted that:
- a dynamic equilibrium will only be achieved in a closed system, (Figure 3.2), that is, none of the components of the equilibrium can escape
- the same equilibrium will be achieved, provided the temperature is constant, no matter from which end it is approached (see Figure 3.3).

## The extent of a chemical reaction

Having established the fact that many reactions do not go to completion but reach a state of dynamic equilibrium, the question arises as to how far the reaction goes before the equilibrium is established. The extent of a reaction when an equilibrium is established is called the **position of equilibrium**. This varies from one reaction to another and depends upon the temperature at which the reaction is performed. If a reaction uses more than 50% of the reactants before reaching equilibrium then the position of equilibrium is said to lie to the right. On the other hand, if less than 50% of reactants are used before the equilibrium is reached, the position of equilibrium is said to lie to the left. It must *not* be assumed that equilibrium is established when there is 50% of reactants and 50% of products.

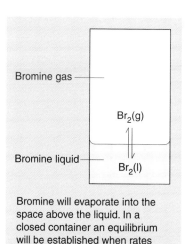

Bromine will evaporate into the space above the liquid. In a closed container an equilibrium will be established when rates of condensation and evaporation are equal.

Fig. 3.2 *Dynamic equilibrium in a closed system*

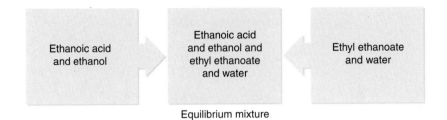

*Fig. 3.3 An equilibrium can be approached from either end*

## The equilibrium constant

The position of equilibrium at a given temperature can be defined by the **equilibrium constant**, which is given the symbol $K_c$ (or $K_p$ for gaseous systems – see Topic 18 in Module 3). This constant relates the equilibrium concentrations, measured in $mol\,dm^{-3}$, of the substances present in the equilibrium mixture.

Thus for the general case of a homogeneous equilibrium involving substances A, B, C, P and Q, for which the equation is:

$$nA + mB + xC \rightleftharpoons pP + yQ$$

the equilibrium constant $K_c$ is given by the expression:

$$K_c = \frac{[P]^p[Q]^y}{[A]^n[B]^m[C]^x}$$

It is important to note that:
- the symbol [ ] represents the mole concentration of a particular species **at equilibrium** in $mol\,dm^{-3}$ and should, strictly speaking, be written $[\ ]_{Equ}$.
- by convention the substances on the right-hand side of the equation are placed in the numerator of the expression
- the concentration of each substance is raised to a power which is the same as the number of moles of that substance in the equation
- all substances present in the equilibrium mixture must be included in the expression for $K_c$.

The value of the equilibrium constant $K_c$ depends *only* on the temperature at which the equilibrium is established. It does not depend on the concentrations of the substances used initially, nor in the equilibrium concentrations which are related to each other and must be such that they maintain $K_c$ constant at a given temperature.

For example, in the reaction:

$$2SO_2(g) + O_2(g) \rightleftharpoons 2SO_3(g)$$

the expression for the equilibrium constant will be:

$$K_c = \frac{[SO_3]^2}{[SO_2]^2[O_2]}$$

## The units of $K_c$

The units of $K_c$ will depend on the actual expression for $K_c$. For example, in the expression for the sulphur dioxide/oxygen/sulphur trioxide equilibrium above, the units for $K_c$ would be:

$$K_c = \frac{(\text{mol dm}^{-3})^2}{(\text{mol dm}^{-3})^2 \times (\text{mol dm}^{-3})} = (\text{mol dm}^{-3})^{-1} = \text{mol}^{-1}\text{dm}^3$$

In the reaction described below, $K_c$ has no units.

## Experimental determination of $K_c$ values

In order to determine the values of $K_c$ by experiment, it is necessary to set up an equilibrium at a given temperature, determine the concentrations of each constituent of the equilibrium mixture, then substitute these values into the expression for the equilibrium constant. As it requires time to determine these concentrations, it is necessary to 'freeze' the equilibrium so that the position of equilibrium does not change while the concentrations are being determined. This is usually done by rapidly reducing the temperature to a level where reaction ceases, in much the same way as can be done in kinetic investigations (Chapter 2). One reaction which is relatively easy to study in the school laboratory is that between a carboxylic acid and an alcohol, for example:

$$CH_3CO_2H(l) + C_2H_5OH(l) \rightleftharpoons CH_3CO_2C_2H_5(l) + H_2O(l)$$

Mixtures of acid and alcohol are heated under reflux, in a thermostatically controlled bath at say 60 °C, for several hours in order to achieve equilibrium. The mixture is rapidly cooled by placing it in a large bath of cold water containing ice. The amount of carboxylic acid remaining can then be determined by titration with standard sodium hydroxide using a suitable indicator such as phenolphthalein. The concentrations of the other substances in the equilibrium mixture need not be experimentally determined since they can be deduced from the stoichiometry of the reaction.

For example, 1 mol of ethanoic acid and 1 mol of ethanol were mixed and heated under reflux at 333 K for several hours until equilibrium was established. The contents of the flask were then poured into some cold water in order to stop any further reaction, and then made up to 1.00 dm$^3$ with distilled water. Portions of this solution, each of 25.0 cm$^3$, required 27.5 cm$^3$ of 0.300 mol dm$^{-3}$ sodium hydroxide for complete neutralisation using phenolphthalein as indicator.

Thus the number of moles of NaOH required = $27.5 \times 0.300/1000$

$$= 8.25 \times 10^{-3}$$

The equation for the reaction between NaOH and ethanoic acid is:

$$CH_3CO_2H(aq) + NaOH(aq) \rightarrow CH_3CO_2^-Na^+(aq) + H_2O(l)$$

Hence the number of moles of ethanoic acid remaining in 25.0 cm$^3$ of the equilibrium mixture is $8.25 \times 10^{-3}$.

The number of moles of ethanoic acid in $1\,dm^3$ of solution

$$= 40 \times 8.25 \times 10^{-3}$$

$$= 0.33\,mol$$

This is therefore the equilibrium concentration of ethanoic acid. Since 1 mol of acid was present initially, 0.67 mol (1 – 0.33) must have been converted into ester and water. Thus, using the stoichiometry of the equilibrium reaction:

$$CH_3CO_2H(l) + C_2H_5OH(l) \rightleftharpoons CH_3CO_2C_2H_5(l) + H_2O(l)$$

the concentrations of ethyl ethanoate and water at equilibrium must each be $0.67\,mol\,dm^{-3}$.

Also, the concentration of ethanol in the equilibrium mixture can be deduced since it reacts with the ethanoic acid in a 1:1 ratio, hence its concentration is also $0.33\,mol\,dm^{-3}$.

All this information can be summarised as shown below:

$$CH_3CO_2H(l) + C_2H_5OH(l) = CH_3CO_2C_2H_5(l) + H_2O(l)$$

| | | | |
|---|---|---|---|
| Initial concentrations (mol dm⁻³): 1.00 | 1.00 | 0 | 0 |
| Equilibrium concentrations (mol dm⁻³): 0.33 | 0.33 | 0.67 | 0.67 |

The equilibrium constant can then be calculated:

$$K_c = \frac{(0.67) \times (0.67)}{(0.33) \times (0.33)} = 4.12$$

$K_c$ has no units in this case.

Other experiments done for the same equilibrium at the same temperature might give the results shown in Table 3.1.

**Table 3.1**

| Initial concentrations /mol dm⁻³ | | Equilibrium concentrations /mol dm⁻³ | | | |
|---|---|---|---|---|---|
| $CH_3CO_2H$ | $C_2H_5OH$ | $CH_3CO_2H$ | $C_2H_5OH$ | $CH_3CO_2C_2H_5$ | $H_2O$ |
| 1.00 | 1.00 | 0.33 | 0.33 | 0.67 | 0.67 |
| 1.00 | 2.00 | 0.15 | 1.15 | 0.85 | 0.85 |
| 2.00 | 2.00 | 0.67 | 0.67 | 1.33 | 1.33 |

The first set of results are those found in the example above and lead to a value of $K_c$ of 4.12. It would be instructive to verify that the other results lead to values of 4.19 and 3.94, respectively, which are in good agreement. Thus whatever the initial quantities, the concentrations of all components present at equilibrium will be such as to maintain a constant value for the equilibrium constant, provided that the temperature is constant.

# Factors affecting the position of equilibrium

There are three factors which can be changed which may result in a change in the position of equilibrium. These are:

- concentration;
- pressure;
- temperature.

**Le Chatelier's principle**: when one of the external factors governing the position of equilibrium is changed, the position of equilibrium will alter in such a way as to oppose the change.

The way in which the position of equilibrium changes, if at all, can be deduced in a qualitative way by applying **Le Chatelier's principle**. Le Chatelier's Principle states that when one of the external factors governing the position of equilibrium is changed, the position of equilibrium will alter in such a way as to oppose the change.

The application of this principle to homogeneous systems only is given in the next three sections.

## The effect of changes in concentration at constant temperature

Consider a system in equilibrium and represented by the equation:

$$n\text{A} + m\text{B} + x\text{C} \rightleftharpoons p\text{P} + y\text{Q}$$

If the concentration of A is increased by adding more A to the system, then the system will react in such a way as to oppose this. Hence a new equilibrium position will be established in which the concentrations of P and Q will be greater than in the original equilibrium. The position of equilibrium has been moved to the right in order to try and reduce the concentration of A.

The same effect would be produced if the concentrations of either B or C were to be increased or if the concentrations of either P or Q were to be reduced by removal from the system.

Similarly, an increase in the concentration of P or Q, or a decrease in the concentration of A, B or C would result in a new equilibrium position being established which would be further to the left. In all cases, the new equilibrium concentrations must be such as to maintain the value of $K_c$ constant.

## The effect of change in pressure at constant temperature

Pressure can only have any effect in a gaseous equilibrium, and then only if there is a change in the total numbers of moles of gas on going from one side of the equilibrium to the other. The number of moles of gas is proportional to the pressure, hence an increase in pressure must result in a new position of equilibrium being established in which there are fewer moles of gas so that the increase in pressure applied is opposed.

For example, for the reaction:

$$2\text{SO}_2(\text{g}) + \text{O}_2(\text{g}) \rightleftharpoons 2\text{SO}_3(\text{g})$$

there are fewer molecules of gas on the right-hand side. An increase in pressure would lead the system to react in such a way as to reduce the total number of molecules of gas and hence reduce the pressure. A new equilibrium position would be established further to the right, that is more $SO_3$ would be formed, and less $SO_2$ and $O_2$ would be present.

In a reaction such as :

$$N_2(g) + O_2(g) \rightleftharpoons 2NO(g)$$

pressure change would have no effect on the equilibrium position since there is no change in the number of moles of gas as the reaction proceeds.

## The effect of change in temperature

In order to predict the effect of temperature changes on the equilibrium position, it is necessary to know whether the enthalpy change for the reaction is positive or negative. Since the reactions are reversible, it must be borne in mind that if $\Delta H$ is positive for the reaction from left to right then it will have the same value with the sign changed for the reaction from right to left.

Thus for any system in equilibrium, for example:

$$2SO_2(g) + O_2(g) \rightleftharpoons 2SO_3(g) \quad \Delta H = -188\,kJ\,mol^{-1}$$

The negative sign for $\Delta H$ is taken to apply to the complete reaction left to right unless stated otherwise. As a result the reaction proceeds to the right with an increase in temperature.

An increase in external temperature would lead to a new position of equilibrium which lies further to the left, that is a new equilibrium mixture would be obtained which contains higher concentrations of $SO_2$ and $O_2$. A decrease in temperature would conversely result in a new equilibrium mixture containing more $SO_3$. This can be summarised as follows:
* exothermic reactions, low temperatures favour the right-hand side;
* endothermic reactions, high temperatures favour the right-hand side.

The reason why temperature affects the position of equilibrium is different from that given for concentration and pressure, because the value of the equilibrium constant changes with temperature. For an exothermic reaction, an increase in temperature would result in a decrease in the equilibrium constant. The position of equilibrium would therefore move to the left, which means that $[SO_3]$ decreases and $[SO_2]$ and $[O_2]$ both increase.

Le Chatelier's principle requires some modification when being applied to heterogeneous systems and will be dealt with in Module 3.

## The rate of attainment of equilibrium

From the discussion of kinetics in Chapter 2 it will be apparent that the factors which affect the position of equilibrium will also affect the rate at which the equilibrium is reached. Thus an increase in concentration, pressure (if gases involved) and temperature will all increase the rate at which the equilibrium is established as well as having an effect on the position of equilibrium. The explanations for this, based on the collision theory, have already been given (see Chapter 2) and need not be repeated here.

A catalyst will increase the rate at which an equilibrium is established but will have no effect on the position of equilibrium. This is because in providing a new route for the reaction, lower activation energy routes are provided for both the forward and the reverse reactions. Hence the rates of both forward and back reactions increase, but the position of equilibrium remains unaltered. Reaction profiles for catalysed and uncatalysed reactions are shown in Figure 3.4.

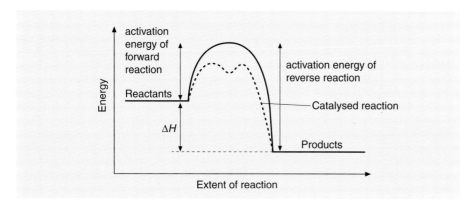

*Fig. 3.4 An energy profile for a reversible reaction which is exothermic in the forward direction, showing the effect of a catalyst*

## Industrial applications

Many industrial processes involve an equilibrium at some point in the manufacturing process. One such is the **Haber–Bosch** process for the manufacture of ammonia.

The key step in this process is the direct synthesis of ammonia from nitrogen and hydrogen, in which an equilibrium is established.

$$N_2(g) + 3H_2(g) \rightleftharpoons 2NH_3(g) \quad \Delta H = -92 \, kJ \, mol^{-1}$$

### Application of the principles of equilibrium

Application of Le Chatelier's principle indicates that the equilibrium in the Haber–Bosch process would move in such a way as to form more $NH_3$ if the temperature were lowered. Also, since there are four moles of gas on the left-hand side and only two moles on the right, an increase in pressure would force the equilibrium to the right. Since the object of the manufacture is to maximise the yield of ammonia, it would seem appropriate to operate at a high pressure and a low temperature. The actual effect of pressure and temperature on the percentage conversion to ammonia is shown in Table 3.2.

As can be seen from Table 3.2, we might predict that use of a pressure of $6.00 \times 10^4 \, kPa$ and a temperature as low as 473 K would achieve almost complete conversion to ammonia.

### Economic factors to be considered

There are, however, other factors which affect the economics of the process and which must be taken into account; these are:

- containers operating at high pressures are expensive to build and to operate
- a reduction in temperature slows down the rate at which the equilibrium is attained
- catalysts can be used in order to increase the rate of reaction
- catalysts are often susceptible to poisoning by impurities present, particularly sulphur compounds and carbon monoxide in the original feedstock
- catalysts usually last longer at lower temperatures
- unreacted gases can be recycled if product can be removed, thus allowing a continuous flow process.

**Table 3.2** *The percentage of nitrogen and hydrogen converted to ammonia for different conditions of temperature and pressure*

| Pressure /$10^2$ kPa | Temperature/K | | | |
|---|---|---|---|---|
| | 473 | 573 | 673 | 773 |
| 10 | 51 | 15 | 4 | 1 |
| 100 | 82 | 53 | 25 | 11 |
| 200 | 89 | 67 | 39 | 18 |
| 300 | 93 | 71 | 47 | 24 |
| 400 | 94 | 80 | 55 | 32 |
| 600 | 95 | 84 | 65 | 42 |

In practice a compromise set of conditions is applied so that the requirement to push the equilibrium position to the right is balanced against the requirement to keep the rate of reaction at a reasonable level. The conditions used in many plants are a pressure of $2.00 \times 10^4$ kPa (200 atm) and a temperature of 673 K (400 °C) which keeps the rate of reaction at a reasonable level as well as giving a reasonably long life to the catalyst (about five years). An iron based catalyst increases the rate of attainment of equilibrium and the incoming gases are purified before entering the catalyst chamber in order to avoid poisoning the catalyst.

Although a conversion of about 40% would be possible under these conditions (as shown in Table 3.2), the gases do not spend long enough in the catalyst chamber to reach equilibrium and a conversion of only about 15% is achieved. The ammonia is cooled in order to liquefy it and remove it from the remaining gases. The unreacted nitrogen and hydrogen are then passed through the catalyst beds again, thus maintaining a continuous circulation.

## The raw materials for the Haber process

The hydrogen is obtained from natural gas, which contains methane, $CH_4$, by reaction with steam.

$$CH_4(g) + H_2O(g) \rightarrow CO(g) + 3H_2(g)$$

Carbon monoxide would poison the catalyst and must therefore be removed from the gases before passing over the catalyst. Air is injected into this mixture and, by a series of reactions, nitrogen and hydrogen are produced in a 1:3 ratio as required by the process and carbon monoxide is removed from the system.

Although natural gas is readily available at the moment, it will not always be so. The alternative fossil fuel, coal, could be used instead, but this is still a finite resource. The equation for the generation of hydrogen would then be:

$$C(s) + H_2O(g) \rightarrow CO(g) + H_2(g)$$

## The manufacture of nitric acid

One of the main uses of ammonia is in the manufacture of nitric acid. The main stage in this process is the oxidation of ammonia:

$$4NH_3(g) + 5O_2(g) \rightleftharpoons 4NO(g) + 6H_2O(g) \quad \Delta H = -909 \text{ kJ mol}^{-1}$$

The equilibrium position would be forced to the left by application of a high pressure and a low temperature, according to Le Chatelier's principle. In practice, however, a low temperature would lead to a reduction in the rate of reaction, hence a compromise temperature is used, together with a catalyst to increase the rate of reaction. Since the volume change is very small in this case (nine moles of gas giving ten moles), the effect of applying pressure is minimal and reasonably low pressures of only 4–10 atmospheres are used. The exact pressure used will depend on economic factors, such as the cost of energy. An excess of air is used as the source of oxygen.

Thus the actual conditions used, leading to about a 96% conversion, are:
- pressure: 4–10 atmospheres;
- temperature: 975–1225 K
- catalyst: platinum containing 10% rhodium.

Care is needed to avoid over-heating the catalyst which would lead to the conversion of ammonia to nitrogen:

$$4NH_3(g) + 3O_2(g) \rightleftharpoons 2N_2(g) + 6H_2O(g) \quad \Delta H = -1636 \text{ kJ mol}^{-1}$$

## Questions

1   State whether the equilibrium position moves to the right, moves to the left or stays the same, for each of the following systems, when the total pressure is increased. State your reasoning in each case.

(a) $N_2(g) + 3H_2(g) \rightleftharpoons 2NH_3(g)$

(b) $H_2(g) + I_2(g) \rightleftharpoons 2HI(g)$

(c) $2SO_2(g) + O_2(g) \rightleftharpoons 2SO_3(g)$

(d) $N_2O_4(g) \rightleftharpoons 2NO_2(g)$

2   For the equilibrium $2NO(g) + O_2(g) \rightleftharpoons 2NO_2(g)$, the enthalpy change for the forward reaction is negative. State and explain the effect on the equilibrium position of each of the following.

(a) An increase in temperature at constant pressure.

(b) A decrease in the concentration of oxygen.

(c) An increase in the total volume of the reaction mixture at constant temperature.

(d) An increase in the concentration of NO.

**3 .** For each of the following equilibria, write the expression for the equilibrium constant, $K_c$ and state its units.

(a) $N_2O_4(g) \rightleftharpoons 2NO_2(g)$

(b) $CH_3CH_2CO_2H(l) + CH_3CH_2OH(l) \rightleftharpoons CH_3CH_2CO_2CH_2CH_3(l) + H_2O(l)$

(c) $H_2(g) + I_2(g) \rightleftharpoons 2HI(g)$

(d) $2SO_2(g) + O_2(g) \rightleftharpoons 2SO_3(g)$

(e) $N_2(g) + 3H_2(g) \rightleftharpoons 2NH_3(g)$

**4** At 723 K, hydrogen gas and iodine gas react together and the following equilibrium is established:

$$H_2(g) + I_2(g) \rightleftharpoons 2HI(g)$$

$K_c$ has a value of 64 for this equilibrium at this temperature. If 3 moles of hydrogen and 3 moles of iodine are mixed in a container of volume 1 dm$^3$, at 723 K, calculate the concentration of the three components when equilibrium is attained.(No knowledge of quadratic equations is required).

**5** When 1.00 mol of ethanoic acid, $CH_3CO_2H$, and 1.00 mol of ethanol, $C_2H_5OH$, were heated in a flask under reflux for several hours an equilibrium was established. The contents of the flask were rapidly cooled and then transferred quantitatively to a graduated flask and made up to 1.000 dm$^3$. Titration of the contents of this flask showed that 0.333 mol dm$^{-3}$ of ethanoic acid were present. Calculate the concentrations of the other components of the equilibrium mixture and hence the value of $K_c$.

What would be the composition of the equilibrium mixture resulting from a similar experiment if 1.00 mol of ethanoic acid, 1.00 mol of ethanol and 1.00 mol of water had been heated to equilibrium at the same temperature?

**6** X is a product of a gaseous reaction which results in an equilibrium mixture being formed:

$$\text{Reactants} \rightleftharpoons \text{X}.$$

The percentage of X in the equilibrium mixture at various temperatures and pressures is shown in the following table.

|        | 1 Atm | 100 Atm | 200 Atm |
|--------|-------|---------|---------|
| 550°C  | 0.77  | 6.70    | 11.9    |
| 650°C  | 0.032 | 3.02    | 5.71    |
| 750°C  | 0.016 | 1.54    | 2.99    |
| 850°C  | 0.009 | 0.87    | 1.68    |

Use this data to deduce, giving your reasoning in each case,

(a) whether the production of X is exothermic or endothermic.

(b) whether the production of X involves an increase or decrease in the number of moles of gas present.

(c) the best conditions to obtain the greatest yield of X.

**7**   For the equilibrium:

$$H_2(g) + I_2(g) \rightleftharpoons 2HI(g) \quad \Delta H = +52 \text{ kJ mol}^{-1}$$

(a) What would be the effect, if any, on this equilibrium of:

   (i)  increasing the total pressure at constant temperature;

   (ii) increasing the temperature at constant pressure?

   Give your reasoning in each case.

(b) The following table shows the concentrations in $\text{mol dm}^{-3}$, of each of the components in the above equilibrium at 600 K.

| Concentration $H_2$ | Concentration $I_2$ | Concentration HI |
|---|---|---|
| $1.71 \times 10^{-3}$ | $2.91 \times 10^{-3}$ | $1.65 \times 10^{-2}$ |

   Calculate the value of $K_c$ at this temperature. What are its units?

**8**   The Haber synthesis of ammonia carried out with an iron catalyst at 450°C and 200 atm pressure, involves the following equilibrium:

$$N_2(g) + 3H_2(g) \rightleftharpoons 2NH_3(g) \quad \Delta H = -92 \text{ kJ mol}^{-1}$$

(a) Why is the synthesis carried out at high pressure?

(b) State one advantage and one disadvantage of performing this reaction at high temperature.

(c) Suggest a reason why the nitrogen and hydrogen must be purified before use.

(d) Assuming that the cost of obtaining the nitrogen and hydrogen gases are fixed, state one factor which would increase significantly the price of ammonia.

(e) Give two commercially important chemicals manufactured from ammonia.

**9**   Ethanol is made industrially by the direct hydration of ethene at 300°C and 70 atm pressure in the presence of phosphoric(V) acid as catalyst:

$$C_2H_4(g) + H_2O(g) \rightleftharpoons C_2H_5OH(g)$$

(a) Explain qualitatively why the synthesis of ethanol is carried out:

   (i)  at increased pressure;

   (ii) at a high temperature in the presence of a catalyst.

(b) State and explain the effect of increasing the proportion of water in the reaction mixture of ethene and water.

# 4 Acid – base equilibria

Fig 4.1 Oranges, lemons and other citrus fruit contain quantities of citric acid

Acids and bases are among the most familiar chemicals in a laboratory and indeed many common household substances contain them. Probably the most common acid is vinegar, which is a very dilute solution of ethanoic acid (often still called acetic acid, $CH_3CO_2H$). Citrus fruits such as oranges, lemons, limes, etc. contain a very weak acid called citric acid, while car batteries contain sulphuric acid. Bases such as ammonia are present in many heavy duty cleaners and sodium hydroxide is to be found in many paint strippers.

Some of these substances can be dangerous if they are spilt on the skin or splashed into the eyes. They need to be treated with caution and appropriate action taken quickly if accidental spillage does occur.

## The Brønsted–Lowry theory

### Definitions

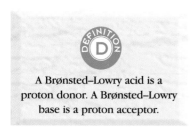

A Brønsted–Lowry acid is a proton donor. A Brønsted–Lowry base is a proton acceptor.

The Brønsted–Lowry theory is based on the definitions that an acid is a proton donor and a base is a proton acceptor. For example, when hydrogen chloride is dissolved in water, the following equilibrium is set up:

$$HCl(g) + H_2O(l) \rightleftharpoons H_3O^+(aq) + Cl^-(aq)$$

In the forward reaction, the HCl is acting as an acid because it donates a proton (an $H^+$ ion) to the water which is acting as a base since it accepts a proton to become $H_3O^+$. In the reverse reaction the $H_3O^+$ ion acts as an acid by donating a proton to $Cl^-$, the latter acting as a base. Thus the equilibrium mixture consists of two acids and two bases and this must always be the case.

## Conjugate acid–base pairs

The equation above can be split into two half-equations which show the proton transfer:

$$HCl - H^+ \rightleftharpoons Cl^-$$
Acid 1          Base 1

$$H_2O + H^+ \rightleftharpoons H_3O^+$$
Base 2          Acid 2

Fig 4.2 Typical household products containing alkalis

This clearly shows that when a species loses a proton, the product has to be a base since the process is reversible. This linkage of an acid to a base by the transfer of a single proton is recognised by use of the word '**conjugate**'. Thus $Cl^-$ is said to be the conjugate base of HCl and HCl the conjugate acid of $Cl^-$. Similarly, $H_3O^+$ and $H_2O$ are a conjugate acid–base pair. Labelling equations as shown above is usually taken to be an adequate indication of the conjugate acid–base link.

Acids which have a single proton to donate are said to be **monoprotic** while those with two or three protons to donate are called **diprotic** and **tripotic**, respectively. Care must be taken with the use of the word 'conjugate' in these cases. For example, for the diprotic acid sulphuric acid ionisation in water occurs in two stages:

$$H_2SO_4 + H_2O \rightleftharpoons H_3O^+ + HSO_4^-$$
Acid 1  Base 2  Acid 2  Base 1

$$HSO_4^- + H_2O \rightleftharpoons H_3O^+ + SO_4^{2-}$$
Acid 3  Base 2  Acid 2  Base 3

The behaviour of the hydrogensulphate ion is different in the two equations, illustrating clearly that 'conjugate' is a relative term and links a specific pair of acids and bases. Thus $HSO_4^-$ is the conjugate base of $H_2SO_4$ but it is the conjugate acid of $SO_4^{2-}$.

*Fig 4.3 Toilet cleaners often contain sodium hydrogensulphate*

Using appropriate molar quantities of sodium hydroxide, two different salts can be formed, $Na_2SO_4$ (sodium sulphate) and $NaHSO_4$ (sodium hydrogensulphate). This latter salt is quite acidic in aqueous solution, as indicated by the second equilibrium above which lies well over to the right. Hence it is used in many powder types of toilet cleaner, its function being to remove lime scale ($CaCO_3$) from the lavatory bowl.

## Acidic solutions

It is important to recognise the difference between the term acid and what is meant by an acidic solution. All too frequently the terms are taken to be synonymous, an acidic solution often being loosely referred to as an acid. For example, the substance HCl is a covalently bonded molecule and acts as an acid as it dissolves in water and donates its proton to the water molecule.

The resulting solution is acidic because it contains the hydrated hydrogen ion which is usually represented as $H_3O^+$ (or, more precisely, the solution is acidic because the concentration of $H_3O^+$ is greater than the concentration of $OH^-$ as will be shown later in this chapter). Such acidic solutions therefore depend on the presence of water as the solvent and they have certain properties in common, such as:
- they react with bases to form salts
- they react with carbonates to produce carbon dioxide gas (the best test for an acidic solution)
- they react with most metals to give hydrogen
- they will produce a certain colour with colour indicators and have a pH less than 7 (see below).

A solution of HCl in a different solvent, such as methylbenzene, would not show any of these acidic properties.

# ACID – BASE EQUILIBRIA

## Bases

The same principles can be applied to the behaviour of bases. Ammonia, for example, is a covalent molecule, but when dissolved in water the following reaction occurs:

$$NH_3(aq) + H_2O(l) \rightleftharpoons NH_4^+(aq) + OH^-(aq)$$

$$\text{Base 1} \qquad \text{Acid 2} \qquad\quad \text{Acid 1} \qquad\quad \text{Base 2}$$

The ammonia is acting as a base by accepting a proton from water. Water is acting as an acid here, in contrast to its behaviour in the previous sections.

Again, the distinction needs to be drawn between a base and an alkaline solution. An alkaline solution is one which contains more $OH^-(aq)$ ions than $H_3O^+(aq)$ ions. Thus the solution formed when ammonia dissolves in water is alkaline because of the $OH^-$ ions produced, and $NH_3$ is acting as a base by accepting a proton.

In all acid–base equilibria the transfer of a proton can only occur if the proton can form a coordinate bond with the species accepting it. Hence the species accepting it must have a lone pair of electrons with can be donated to form this coordinate bond.

This forms the basis of another theory of acids and bases, but which is beyond the scope of the syllabus being studied.

## The effect of solvent on acid–base behaviour

In solvents other than water the behaviour of substances which are normally regarded as acids can be considerably modified. For example, ethanoic acid reacts with water, giving an acidic solution:

$$CH_3CO_2H(l) + H_2O(l) \rightleftharpoons CH_3CO_2^-(aq) + H_3O^+(aq)$$

$$\text{Acid 1} \qquad\quad \text{Base 2} \qquad\qquad \text{Base 1} \qquad\qquad \text{Acid 2}$$

If, however, the ethanoic acid were to be dissolved in concentrated HCl the following equilibrium would be set up, in which the ethanoic acid behaves as a base.

$$CH_3CO_2H + HCl \rightleftharpoons CH_3CO_2H_2^+ + Cl^-$$

$$\text{Base 1} \qquad\quad \text{Acid 2} \qquad\quad \text{Acid 1} \qquad\quad \text{Base 2}$$

Similarly, a mixture of concentrated nitric acid and concentrated sulphuric acid leads to some very surprising behaviour, with the concentrated nitric acid acting as a base:

$$HNO_3 + H_2SO_4 \rightleftharpoons H_2NO_3^+ + HSO_4^-$$

$$\text{Base 1} \qquad\quad \text{Acid 2} \qquad\quad \text{Acid 1} \qquad\quad \text{Base 2}$$

# Strengths of acids

## Strong and weak acids

The 'strength' of an acid is a term used to indicate the amount of ionisation which occurs when the acid is dissolved in water. In the case of HCl in water, HCl is in fact a covalent molecule which forms ions when reacted with water:

$$HCl(g) + H_2O(l) \rightleftharpoons H_3O^+(aq) + Cl^-(aq)$$

In this particular case the equilibrium lies so far to the right that the acid can be considered to be completely ionised and it is said to be a strong acid. Other examples of strong acids are $HNO_3$ and (for its first ionisation) $H_2SO_4$, all of which are assumed to be fully ionised in aqueous solution.

With acid such as ethanoic acid, the equilibrium:

$$CH_3CO_2H(l) + H_2O(l) \rightleftharpoons CH_3CO_2^-(aq) + H_3O^+(aq)$$

lies much further to the left and the solution will contain $CH_3CO_2H$ molecules as well as $CH_3CO_2^-(aq)$ and $H_3O^+(aq)$ ions. Ethanoic acid is therefore designated a 'weak' acid. The terms 'strong' and 'weak' are obviously only comparative. It is difficult to know where to draw the line between them since acids can range from being 1 or 2% ionised to 100% ionised. A quantitative measure of acid strength is discussed in the next section.

Similarly, there are strong and weak bases. The hydroxides of the metals of groups 1 and 2 of the Periodic Table are strong and are considered to be 100% ionised in aqueous solution. Ammonia, however, is a weak base because the ionisation in water is small, i.e. the position of the equilibrium:

$$NH_3(aq) + H_2O(l) \rightleftharpoons NH_4^+(aq) + OH^-(aq)$$

lies well over to the left. There are, however, sufficient hydroxide ions present to precipitate many metal hydroxides from solutions of the metal salt. This was demonstrated in Module 1.

## Evidence from enthalpies of neutralisation

The enthalpy of neutralisation is defined as the enthalpy change when one mole of water is formed from reaction of an acid with a base. For example, it would be $\Delta H$ for any of the following reactions:

$$HCl(aq) + NaOH(aq) \rightarrow NaCl(aq) + H_2O(l)$$

$$\tfrac{1}{2}H_2SO_4(aq) + NaOH(aq) \rightarrow \tfrac{1}{2}Na_2SO_4(aq) + H_2O(l)$$

$$CH_3CO_2H(aq) + NaOH(aq) \rightarrow CH_3CO_2Na(aq) + H_2O(l)$$

A method for its measurement in the laboratory was given in Chapter 1. The value of the enthalpy of neutralisation is remarkably constant at $-56\,kJ\,mol^{-1}$ when the acid and the base are both strong. This is not really surprising when it is realised that if the acid and base are fully ionised, the ionic equation for all these reactions is the same, that is:

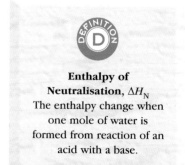

**Enthalpy of Neutralisation, $\Delta H_N$**
The enthalpy change when one mole of water is formed from reaction of an acid with a base.

$$H^+(aq) + OH^-(aq) \rightarrow H_2O(l)$$

Hence we are simply measuring the enthalpy change for the same reaction. If either the acid or base is weak, however, then some of the energy is absorbed in order to ionise the weak acid or base. Hence the amount of heat liberated is usually less than $-57 \, kJ \, mol^{-1}$.

For example, the enthalpy of neutralisation for ethanoic acid and sodium hydroxide is $-55.2 \, kJ \, mol^{-1}$. Assuming the base to be fully ionised, the following diagram illustrates the enthalpy changes occurring:

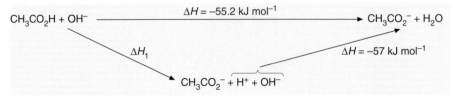

Application of Hess's law enables us to calculate the amount of energy ($\Delta H_1$) which has been required to ionise the weak acid :

$$-55.2 = \Delta H_1 + (-57)$$
$$\therefore \Delta H_1 = 57 - 55.2 = +1.8 \, kJ \, mol^{-1}$$

This is not a true figure for the full amount of energy required to ionise the acid since there will also be heat liberated on hydration of the ions once they have been formed. The value of $+1.8 \, kJ \, mol^{-1}$ is the sum of both these processes. Energy released on hydration of the ions is sometimes greater than the energy required to complete the ionisation of the acid or base, and this explains why the values of enthalpy of neutralisation of some acids and bases are greater than $-57 \, kJ \, mol^{-1}$. Some values for other acids and bases are shown in Table 4.1.

**Table 4.1** *The enthalpies of neutralisation of some acids and bases.*

| Reaction | $-\Delta H/kJ \, mol^{-1}$ |
|---|---|
| HCl/NaOH | 57.1 |
| $HNO_3$/KOH | 57.3 |
| $CH_3CO_2H$/NaOH | 55.2 |
| $HCl/NH_3$ | 52.2 |
| $HNO_3/^1/_2Ba(OH)_2$ | 58.2 |

## The acid dissociation constant

The strength of an acid can be measured in a quantitative way by measuring the equilibrium constant which determines the position of equilibrium. Consider, for example, a monoprotic acid HA which dissociates in water:

$$HA(aq) + H_2O(l) \rightleftharpoons H_3O^+(aq) + A^-(aq)$$

The equilibrium constant, $K_c$, for this would normally be written:

$$K_c = \frac{[H_3O^+][A^-]}{[HA][H_2O]}$$

Since the concentration of $H_2O(l)$ is effectively constant, it can be incorporated into the value for $K_c$ so that it becomes a new constant which is given the symbol $K_a$ where:

$$K_a = \frac{[H_3O^+][A^-]}{[HA]}$$

$K_a$ is known as the **acid dissociation constant**.

The following points should be noted.
- the expression must not have $[H_2O]$ on the bottom line
- $K_a$ is essentially an equilibrium constant, it will be temperature dependent
- $K_a$ will have units, in this case $mol\,dm^{-3}$
- The value of $K_a$ will determine the position of equilibrium. The greater the value of $K_a$, the stronger the acid HA

Thus a quantitative comparison of the strengths of acids *when dissolved in water*, can be made. It is important to realise that these values of $K_a$ are only valid for solutions in water where the base is the same (i.e. $H_2O$) no matter which acid is dissolved in it. The strengths may be different in other solvents.

Some values of $K_a$ for some common acids are given in Table 4.2.

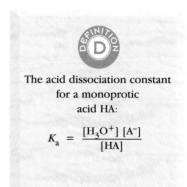

The acid dissociation constant for a monoprotic acid HA:

$$K_a = \frac{[H_3O^+][A^-]}{[HA]}$$

**Table 4.2** *The $K_a$ values for some common weak acids*

| Acid | $K_a/mol\,dm^{-3}$ |
|---|---|
| ethanoic acid | $1.70 \times 10^{-5}$ |
| chloroethanoic acid | $1.38 \times 10^{-3}$ |
| dichloroethanoic acid | $5.13 \times 10^{-2}$ |
| benzenecarboxylic acid | $6.30 \times 10^{-5}$ |
| nitrous acid | $5.00 \times 10^{-4}$ |

For polyprotic acids, a $K_a$ value can be written for each stage of the ionisation and these show that the ionisation gets progressively weaker. The values for phosphoric(V) acid are shown in Table 4.3.

**Table 4.3** *$K_a$ values for the successive ionisations of phosphoric(V) acid*

| $K_a/mol\,dm^{-3}$ | $K_a/mol\,dm^{-3}$ |
|---|---|
| $H_3PO_4 + H_2O \rightleftharpoons H_3O^+ + H_2PO_4^-$ | $8.0 \times 10^{-3}$ |
| $H_2PO_4^- + H_2O \rightleftharpoons H_3O^+ + HPO_4^{2-}$ | $6.3 \times 10^{-8}$ |
| $HPO_4^{2-} + H_2O \rightleftharpoons H_3O^+ + PO_4^{3-}$ | $4.0 \times 10^{-13}$ |

## Ionisation of water

Water, no matter how pure it is made, always ionises to a very small extent:

$$H_2O(l) + H_2O(l) \rightleftharpoons H_3O^+(aq) + OH^-(aq)$$

and this is itself an acid–base equilibrium. This ionisation is particularly interesting in that the water produces its own conjugate acid and its own conjugate base at the same time.

Application of the equilibrium law to this leads to the following expression for the equilibrium constant $K_c$:

$$K_c = \frac{[H_3O^+(aq)][OH^-(aq)]}{[H_2O]^2}$$

The concentration of water molecules $[H_2O]^2$ is effectively constant and hence the top line of the above expression must also be constant. Hence:

$$[H_3O^+(aq)][OH^-(aq)] = \text{constant} = K_w$$

$K_w$ is called the **ionic product of water**.

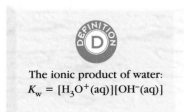

The ionic product of water:
$K_w = [H_3O^+(aq)][OH^-(aq)]$

Since $K_w$ is an equilibrium constant, it is temperature dependent. $K_w$ will have units of $mol^2\,dm^{-6}$. Its value is about $1 \times 10^{-14}$ at 298 K, but the exact value will vary with temperature.

In water itself the concentration of the $H_3O^+$ ion must be the same as that of the $OH^-$ ion. Thus:

$$[H_3O^+(aq)] = [OH^-(aq)] = \sqrt{1 \times 10^{-14}} = 1 \times 10^{-7}\,mol\,dm^{-3}$$

When acids or alkalis are dissolved in water there will be different concentrations of $H_3O^+(aq)$ and $OH^-(aq)$ but the product of these two concentrations must always be equal to the value of $K_w$ at the appropriate temperature.

If, for example, a solution is made which contains $0.1\,mol\,dm^{-3}$ HCl, this will provide a concentration of $H_3O^+(aq)$ of $0.1\,mol\,dm^{-3}$ (assuming that the acid ionises completely). Thus the water ionisation is suppressed, i.e. the equilibrium moves to the left, but the value of $K_w$ must still be maintained. Assuming that the $[H_3O^+(aq)]$ from the water is now so small compared to that from the acid that it is negligible:

$$[H_3O^+(aq)][OH^-(aq)] = K_w = 10^{-14}$$

$$\therefore 0.1 \times [OH^-] = 10^{-14}$$

$$\therefore [OH^-] = 10^{-13}\,mol\,dm^{-3}$$

There is, therefore, a very small, but definite, concentration of hydroxide ions in a solution of an acid. Application of the same principles leads to the conclusion that solutions of alkalis have a small, but definite, hydrogen ion concentration.

# Measurement of acidity – the pH scale

The acidity of a solution depends on the concentration of $H_3O^+$ ions and is measured on the pH scale, pH being defined as :

$$pH = -\log_{10} [H_3O^+]$$

The use of this scale does do away with awkward numbers, particularly negative powers of ten. However, it must be noted that:
- the negative sign in the definition does mean that the pH decreases as the hydrogen ion concentration increases
- a change of 1 unit on the pH scale corresponds to an tenfold change in the hydrogen ion concentration (since a $\log_{10}$ scale is used).

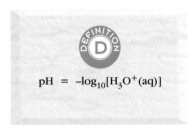

$$pH = -\log_{10}[H_3O^+(aq)]$$

Thus if the hydrogen ion concentration in a solution is decreased ten times, the pH will increase by one unit.

The mathematical relationship between pH and the hydrogen ion concentration is such that if the pH is a whole number, then the conversion to the corresponding hydrogen ion concentration is very simple. If the pH is $x$, then the hydrogen ion concentration is $10^{-x}\,mol\,dm^{-3}$. The relationship between the two quantities is shown in the following scale:

| pH | 0 | 1 | 2 | 3 | 4 | 5 | 6 | 7 | 8 | 9 | 10 | 11 | 12 | 13 | 14 |
|---|---|---|---|---|---|---|---|---|---|---|---|---|---|---|---|
| $[H_3O^+]$ /mol dm$^{-3}$ | 1 | $10^{-1}$ | $10^{-2}$ | $10^{-3}$ | $10^{-4}$ | $10^{-5}$ | $10^{-6}$ | $10^{-7}$ | $10^{-8}$ | $10^{-9}$ | $10^{-10}$ | $10^{-11}$ | $10^{-12}$ | $10^{-13}$ | $10^{-14}$ |

This is the range of pH which is most commonly used. However, pH values of less than 0 and values of greater than 14 are possible.

As shown above, the hydrogen ion concentration in pure water is $1 \times 10^{-7}\,mol\,dm^{-3}$, hence pure water has a pH of 7. Since pure water is the end product of the neutralisation of a strong acid and a strong base, a pH of 7 is regarded as defining what is meant by a neutral solution at 25°C. Solutions with a pH of less than 7 are therefore regarded as acidic and those with a pH greater than 7 are alkaline.

## Calculation of the pH of strong acids

For purposes of calculation, it is assumed that all strong acids are completely ionised. Hence the hydrogen ion concentration is obtained directly from the molarity of the acid. For example, in a solution of hydrochloric acid which is $0.1\,mol\,dm^{-3}$, the acid ionises completely:

$$HCl(aq) + H_2O(l) \rightarrow H_3O^+(aq) + Cl^-(aq)$$

so, $[H_3O^+] = 0.1\,mol\,dm^{-3}$     $\log_{10} 0.1 = -1$

$\therefore -\log_{10} 0.1 = 1$     $\therefore pH = 1$

With diprotic acids such as sulphuric acid $H_2SO_4$ it might be thought that of the hydrogen ion concentration is double that the acid, so that a $0.1\,mol\,dm^{-3}$ solution of sulphuric acid would have $[H_3O^+] = 0.2\,mol\,dm^{-3}$. This is not so. The second ionisation of sulphuric acid

$$HSO_4^- (aq) + H_2O(aq) \rightleftharpoons SO_4^{2-}(aq) + H_3O^+(aq)$$

is fairly weak, with $K_a = 0.01$ mol dm$^{-3}$. It is not very difficult to show that this contributes so little to the hydrogen ion concentration that the pH of 0.1 mol dm$^{-3}$ sulphuric acid is about 0.98, very little different from HCl (aq) of the same concentration with pH = 1.

## Calculation of the pH of strong bases

Strong bases are also assumed to be completely ionised. The hydroxide ion concentration is therefore easily obtained from the molarity of the base. For example a 0.3 mol dm$^{-3}$ solution of sodium hydroxide is ionised:

$$NaOH(aq) \rightarrow Na^+(aq) + OH^-(aq)$$

So, $[OH^-] = 0.3$ mol dm$^{-3}$

The hydrogen ion concentration which is present in this solution is determined by the value of $K_w$ for water. If this is taken to be $10^{-14}$ mol$^2$ dm$^{-6}$ at 298 K, then the hydrogen ion concentration at this temperature is given by:

$$[H_3O^+][OH^-] = 10^{-14}$$

$$\therefore 0.3 \times [H_3O^+] = 10^{-14}$$

$$[H_3O^+] = 10^{-14} \div 0.3 = 3.33 \times 10^{-14}$$

$$\therefore pH = -\log_{10} 3.33 \times 10^{-14} = 13.5$$

## Calculation of the pH of weak acids

More information is required to calculate the pH of weak acids. As well as the molarity of the solution, it is necessary to know the degree of ionisation of the acid, or its $K_a$ value. For example, calculate the pH of a 0.1 mol dm$^{-3}$ solution of ethanoic acid at 298 K given that $K_a = 1.7 \times 10^{-5}$ at this temperature. The equilibrium

$$CH_3CO_2H(aq) + H_2O(l) \rightleftharpoons CH_3CO_2^-(aq) + H_3O^+(aq)$$

is often represented in a simpler form for the purposes of calculation:

$$CH_3CO_2H(aq) \rightleftharpoons CH_3CO_2^-(aq) + H^+(aq)$$

$$K_a = \frac{[CH_3CO_2^-(aq)][H^+(aq)]}{[CH_3CO_2H(aq)]}$$

The ethanoate ions and hydrogen ions must be produced in equal concentrations, so:

$$K_a = \frac{[H^+(aq)]^2}{[CH_3CO_2H(aq)]}$$

For a weak acid, the ionisation is assumed to be so small that it is negligible. Hence $[CH_3CO_2H(aq)]$ is assumed to be 0.1 mol dm$^{-3}$.

$$\therefore K_a = 1.7 \times 10^{-5} = \frac{[H^+(aq)]^2}{0.1}$$

$$[H^+(aq)]^2 = 0.1 \times 1.7 \times 10^{-5}$$
$$[H^+(aq)] = \sqrt{0.1 \times 1.7 \times 10^{-5}} = 1.3 \times 10^{-3}\,mol\,dm^{-3}$$
$$\therefore pH = -\log_{10} 1.3 \times 10^{-3} = 2.88$$

In the above calculations it is important that you are able to carry out the calculations in reverse, that is, given the pH value you should be able to calculate the hydrogen ion concentration for any strong acid or base or calculate the $K_a$ value for a weak acid.

For example, if a solution of hydrochloric acid has a pH of 3, calculate the hydrogen ion concentration and the molarity of the acid.

$$pH = 3 \therefore [H_3O^+] = antilog\,(-3) = 1 \times 10^{-3}\,mol\,dm^{-3}$$

Assuming the acid to be 100% ionised, and since the acid is monoprotic, the molarity of the acid must also be $1 \times 10^{-3}\,mol\,dm^{-3}$.

If a solution of a strong diprotic acid $H_2SO_4$ also has a pH of 3, then its hydrogen ion concentration will also be $1 \times 10^{-3}\,mol\,dm^{-3}$. However this comes from two ionisations, the first being strong and the second weak. The number of moles of sulphuric acid required is fewer than for hydrochloric acid, but because the second ionisation of sulphuric acid is weak the required concentration is not halved, which might have been expected from its diprotic nature.

It is important to appreciate the following points.
1 The difference between the strength of an acid and the concentration of the solution.

Strength refers to the extent of ionisation or dissociation of the acid into hydrogen ions, while concentration refers to the number of moles of acid dissolved in a given volume of water. It is therefore quite possible to have a concentrated solution of a weak acid and a dilute solution of a strong acid.

2 The pH value of a solution by itself is not a measure of the strength of an acid.

The pH of a solution is simply related to the hydrogen ion concentration in the solution. Some knowledge of the molar concentration of the acid solution is needed before a judgement can be made about the strength of the acid from its pH value. It was shown in an example above that the pH of $0.1\,mol\,dm^{-3}$ solution of ethanoic acid is 2.88. A $0.001\,mol\,dm^{-3}$ solution of hydrochloric acid would have a pH of 3.0. This does not mean that ethanoic acid is a stronger acid than hydrochloric acid, but simply that, for the concentrations of solution given, it has a greater hydrogen ion concentration and hence a lower pH value.

## Buffer solutions

### The nature and function of buffer solutions
Buffer solutions are solutions of known pH which have the ability to resist changes in pH when contaminated by small amounts of acid or alkali. Buffer solutions are very important in certain situations where the pH of a solution

**Buffer solutions** are solutions of known pH which have the ability to resist changes in pH when contaminated with small amounts of acid or alkali.

must be maintained. Blood, for example, is buffered at pH 7.4 and a variation of only 0.5 in this pH could prove fatal. It is obviously important that any injections into the bloodstream, given for medical reasons, should also be buffered.

The simplest form of buffer solution is made using a solution containing a weak acid together with a salt of the same acid, for example, ethanoic acid plus sodium ethanoate. The salt obviously has an important function which can be understood if we first consider what would happen in a solution of the acid on its own.

A weak acid HA will dissociate in solution as follows:

$$HA(aq) + H_2O(l) = H_3O^+(aq) + A^-(aq)$$

where $[H_3O^+(aq)] = [A^-(aq)]$ and both concentrations are small. The HA could cope quite well with the addition of $OH^-$ ions since they would combine with the $H_3O^+$ to form water. The equilibrium would move to the right to replace the $H_3O^+$ removed and hence the pH would be maintained. Addition of more $H_3O^+$ ions however, would push the equilibrium to the left by combining with $A^-$ ions. The small concentration of $A^-$ ions would very rapidly fall to zero, any further addition of $H_3O^+$ would not be removed and hence the pH would change rapidly.

The presence of the salt of the weak acid $Na^+A^-$ gives an additional supply of $A^-$ ions and hence the reaction with added $H_3O^+$ ion can continue for much longer and a change in pH is thus resisted. Another function of the salt is that the added $A^-$ ions suppress the already weak ionisation of HA. Thus [HA] is actually greater than it is in the acid itself and there is an even greater reservoir of HA molecules to remove any added $OH^-$ ions.

The buffer solution acts as follows:
 • Added hydrogen ions are removed from solution by:

$$H_3O^+(aq) + A^-(aq) \rightleftharpoons HA(aq) + H_2O(l)$$

 • Added hydroxide ions are removed from solution by the reaction:

$$HA(aq) + OH^-(aq)) \rightleftharpoons H_2O(l) + A^-(aq)$$

During these processes, neither [HA] nor $[A^-]$ change by very much and so the hydrogen ion concentration does not change very much either: the pH remains almost constant.

## Calculation of the pH of a buffer solution

Consider a solution at 298 K which contains $0.1\,mol\,dm^{-3}$ ethanoic acid and $0.2\,mol\,dm^{-3}$ sodium ethanoate, the $K_a$ value for ethanoic acid being $1.7 \times 10^{-5}\,mol\,dm^{-3}$ at this temperature.

The calculation is made from the $K_a$ expression:

$$K_a = \frac{[CH_3CO_2^-(aq)][H^+(aq)]}{[CH_3CO_2H(aq)]}$$

but in this situation, $[CH_3CO_2^-(aq)]$ and $[H^+(aq)]$ are *not* equal, since the salt has been added to the solution.

Two assumptions are made:

- The anion concentration from the acid is very small compared to that from the salt (which is assumed to be 100% ionised). The salt is considered to be the only source of the anion. So, in this case:

$$[CH_3CO_2^-(aq)] = [\text{sodium ethanoate}] = 0.2 \, mol \, dm^{-3}$$

- The presence of excess anion from the salt has suppressed the ionisation of the acid to such a low level that it can be considered negligible. So, in this case:

$$[CH_3CO_2H(aq)] = [\text{ethanoic acid}] = 0.1 \, mol \, dm^{-3}$$

$$1.7 \times 10^{-5} = \frac{[H_3O^+] \times 0.2}{0.1}$$

$$[H_3O^+] = \frac{1.7 \times 10^{-5} \times 0.1}{0.2} = 8.5 \times 10^{-6} \, mol \, dm^{-3}$$

$$pH = -\log_{10} 8.5 \times 10^{-6} = 5.07$$

The effect on this buffer solution of adding hydrogen (or hydroxide) ions can also be calculated. Suppose, for example, that $1 \, cm^3$ of $1.0 \, mol \, dm^{-3}$ hydrochloric acid is added to $1 \, dm^3$ of the above buffer solution. The hydrogen ion concentration has been increased by $10^{-3} \, mol \, dm^{-3}$. This is removed by:

$$H_3O^+(aq) + A^-(aq) = HA(aq) + H_2O$$

The concentration of $A^-$ will thus decrease by 0.001 to a new value of $0.2 - 0.001 = 0.199 \, mol \, dm^{-3}$.
The concentration of HA will increase to $0.1 + 0.001 = 0.101 \, mol \, dm^{-3}$.
The concentration of hydrogen ions which can co-exist with these new concentrations is calculated from:

$$[H_3O^+] = \frac{1.7 \times 10^{-5} \times 0.101}{0.199} = 8.63 \times 10^{-6} \, mol \, dm^{-3}$$

$$\therefore pH = -\log_{10} 8.63 \times 10^{-6} = 5.06$$

The pH has changed by only 0.01 unit.

The addition of the same amount of acid to $1 \, dm^3$ of distilled water would reduce the pH from 7 to 3.

## Amino acids as buffer solutions

Amino acids of the type represented by the general formula $R-CH(NH_2)CO_2H$ are very important biologically. They are able to act as buffer solutions since they have a $-NH_2$ group which can act as a base and combine with protons to form $-NH_3^+$. They also contain a $-CO_2H$ group which can react with hydroxide ions by loss of a proton forming $-CO_2^-$. In fact the amino acid normally exists as a zwitterion: $R-CH(NH_3^+)CO_2^-$, the $-CO_2H$ donating a proton to $-NH_2$.

Addition of acid to the zwitterion leads to protons being removed:

$$R–CH(NH_3^+)CO_2^- + H^+ \rightarrow R–CH(NH_3^+)CO_2H$$

Addition of alkali leads to $OH^-$ being removed:

$$R–CH(NH_3^+)CO_2^- + OH^- \rightarrow R–CH(NH_2)CO_2^- + H_2O.$$

## Acid–base titrations

A titration is a quantitative technique in which the volume of one solution required to completely react with a known volume of another solution is accurately measured. This requires us to have some means of knowing when the reaction is complete, that is, when the acid and base are present in the stoichiometric proportions as shown by the equation. This is called the **end point** or **equivalence point** for the titration.

Certain colour indicators, which have different colours depending on the pH of the solution in which they are placed, can be used.  In order to understand how these colour indicators work, it is necessary first of all to understand the way in which the pH changes when an acid is gradually added to a base, or vice versa, until in excess.

### pH changes during titrations

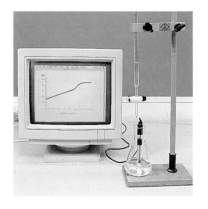

Fig 4.4 Producing a pH curve from an acid–base titration

The change in pH during a titration can be followed by inserting a pH electrode into the solution and measuring the pH after each addition of acid or base. The graph of pH against volume of acid or base added is then plotted to give a so-called titration curve. The shape of the curve obtained is dependent to some extent on the strengths of the acid and base used, and these are shown in Figure 4.5 for various combinations of strong and weak acids and bases.

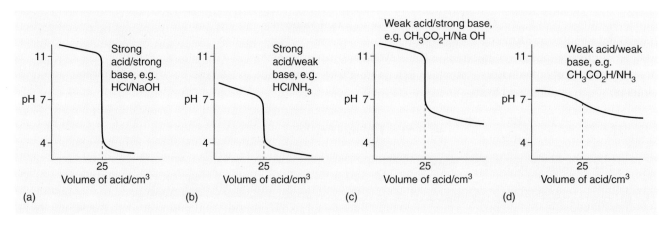

Fig. 4.5 Titration curves for various acids and bases. Each curve relates to the volume of $0.10\,mol\,dm^{-3}$ acid added to $25\,cm^3$ of $0.10\,mol\,dm^{-3}$ base

It should be noted that the amount of acid required to neutralise the base in all these cases does **not** depend on the strength of the acid but on the stoichiometric equation for the reaction.

## Buffering effects during titrations

As can be seen in Figure 4.5 there is a large change in pH for a very small addition of acid around the end point of the titration, in all cases except that for weak acid/weak base. The pH change is about 7 units for strong acid/strong base titrations but only about 4 units for weak acid/strong base and for strong acid/weak base. The change is not so great in these cases since the solution just before or after the end point consists of a buffer solution which resists such large changes in pH.

In the case of the titration of $CH_3CO_2H$ with NaOH, the solution after the end point contains sodium ethanoate and ethanoic acid which is a buffer solution, as described above. Similarly, in the case of HCl reacting with $NH_3$, the solution just before the end point consists of the weak base ammonia and its salt ammonium chloride. Hence, the solution is again buffered.

Weak acid/weak base titrations such as $CH_3CO_2H$ with $NH_3$, contain a buffer solution both before the end point ($NH_3$ + ammonium ethanoate) as well as after the end point ($CH_3CO_2H$ + ammonium ethanoate), hence the large change in pH does not occur.

## The pH at the end point

It can be seen from Figure 4.1 that the pH at the end point of a titration is not always 7. In the case of a weak acid /strong base titration such as $CH_3CO_2H$ with NaOH, the pH at the end point is greater than 7. The only substances present in the solution at the end point are the salt sodium ethanoate, and water. The anion is the conjugate base of a weak acid and hence a reasonably strong base, capable of reacting with water. This is known as salt hydrolysis:

$$CH_3CO_2^- \text{ (aq)} + H_2O(l) \rightleftharpoons CH_3CO_2H(aq) + OH^-(aq)$$

In this solution, $[OH^-] > [H_3O^+]$ and the solution has a pH greater than 7. Similarly, the solution at the end point for the titration of ammonia and hydrochloric acid should contain ammonium chloride and water only. The ammonium ion, however, will undergo hydrolysis, producing hydrogen ions:

$$NH_4^+(aq) + H_2O(l) \rightleftharpoons NH_3(aq) + H_3O^+(aq)$$

The $NH_4^+$ is the conjugate acid of a weak base ($NH_3$) and is therefore capable of acting as a reasonably strong acid. As a result, the equilibrium lies to the right and the solution is acidic and its pH is less than 7.

# Colour indicators

## Behaviour of colour indicators

Colour indicators are complex organic molecules, but their behaviour is relatively simple to understand. They are all very weak acids in which the conjugate base is a different colour from the acid itself. They can be represented by the equilibrium:

$$HIn + H_2O \rightleftharpoons In^- + H_3O^+$$

Colour 1            Colour 2

The strengths of indicators as acids will depend on the $K_a$ value:

$$K_a = \frac{[In^-]\,[H_3O^+]}{[HIn]}$$

Addition of acid to this solution will push the equilibrium to the left and colour 1 will be seen. Addition of alkali will result in the equilibrium moving to the right since the $OH^-$ ions will remove hydrogen ions from the equilibrium. Colour 2 will be seen. The indicator will show an intermediate colour when colour 1 and colour 2 are present in equal concentrations i.e. when $[In^-]$ = $[HIn]$. In this situation:

$$K_a = \frac{[In^-]\,[H_3O^+]}{[HIn]} \quad \text{and therefore} \quad K_a = [H_3O^+]$$

The indicator will show its intermediate colour at a given pH value which is determined by the $K_a$ value for the indicator. Since all indicators have different $K_a$ values, they will change colour at different pH values. The complete colour change from colour 1 to colour 2, or vice versa, requires a change of about 1.5 to 2 units of pH. This range is known as the **working range** of the indicator. The pH of the intermediate colour is at the midpoint of this range. Some values for common indicators are shown in Table 4.4.

**Table 4.4** *The Working Range and colour changes of some Indicators*

| Indicator | pH range | Acid colour | Alkaline colour |
|---|---|---|---|
| methyl orange | 3.2–4.5 | red | yellow |
| phenolphthalein | 8.2–10.0 | colourless | magenta |
| bromothymol blue | 6.0–7.0 | yellow | blue |

*Fig 4.6 The colouring material in red cabbage can be used as an indicator. The red coloured extract shown in the left-hand tube is yellow with acid (right-hand tube) and blue/purple with alkali (middle tube)*

## Choice of indicator for titrations

The colour indicator used for any acid–alkali titration should ideally change colour at the pH corresponding to the midpoint of the straight portion of the titration curve. However, there is no serious loss of accuracy if the indicator changes colour anywhere on the straight portion of the titration curve since the pH change is so large for the addition of a very small amount of acid or base. For a strong acid/strong base titration, any of the indicators in Table 4.4 could be used. More care is needed, however, in the choice of indicator when either the acid or base is weak since the pH range of the straight portion is smaller. Thus phenolphthalein could be used for strong base/weak acid, whereas methyl orange could not. Similarly methyl orange would be much more suitable than phenolphthalein for strong acid/weak base titrations.

## Titration curves for diprotic acids

A diprotic acid such as ethanedioic acid $H_2C_2O_4$ has two protons which can be successively replaced in what are effectively separate acid–base reactions.

$$H_2C_2O_4(aq) \; + \; NaOH(aq) \rightarrow NaHC_2O_4(aq) \; + \; H_2O(l) \qquad \text{Step 1}$$

$$NaHC_2O_4(aq) \; + \; NaOH(aq) \rightarrow Na_2C_2O_4(aq) \; + \; H_2O(l) \qquad \text{Step 2}$$

The titration curve for such a reaction shows two 'kinks', each corresponding to the completion of one of these steps, as shown in Figure 4.2.

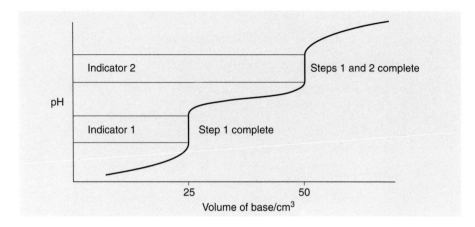

*Fig. 4.7 Titration curve for the titration of 25 cm³ 0.10 mol dm⁻³ ethanedioic acid with 0.10 mol dm⁻³ sodium hydroxide*

An appropriate choice of indicator would allow either of the two end points to be identified. It is obviously important to know which end point has been identified when using the results of such titrations for calculation, since the equation used in the calculation must be the correct one for the indicator used.

Thus if indicator 1 is used, the equation for step 1 only is appropriate. If indicator 2 is used then the equations for step 1 and step 2 must be added together.

## Questions

1   (a) Define what is meant by an acid according to the Brønsted–Lowry Theory.

(b) Identify the conjugate acid–base pairs in the following equilibria.

$$CH_3CO_2H + H_2O \rightleftharpoons CH_3CO_2^- + H_3O^+$$

$$CH_3CO_2H + HCl \rightleftharpoons CH_3CO_2H_2^+ + Cl^-$$

$$CH_3NH_2 + H_2O \rightleftharpoons CH_3NH_3^+ + OH^-$$

$$2H_2O \rightleftharpoons H_3O^+ + OH^-$$

2   (a) Calculate the hydrogen ion concentration of a solution which contains 0.100 mol dm⁻³ of ethanoic acid, the $K_a$ value being $1.8 \times 10^{-5}$ mol dm⁻³ at 298 K.

(b) Calculate the pH of this solution.

3 Calculate the pH of the following, taking the value of $K_w$ to be $1 \times 10^{-14}\,mol\,dm^{-3}$, where appropriate:

(a) hydrochloric acid of concentration $0.001\,mol\,dm^{-3}$;

(b) sulphuric acid of concentration $0.001\,mol\,dm^{-3}$;

(c) potassium hydroxide of concentration $0.002\,mol\,dm^{-3}$;

(d) benzenecarboxylic acid of concentration $0.1\,mol\,dm^{-3}$, $K_a = 6.3 \times 10^{-5}\,mol\,dm^{-3}$.

4 (a) Calculate the concentration of the acid in each of the following solutions, both of which have a pH of 3:

(i) hydrochloric acid;

(ii) methanoic acid which is a weak monoprotic acid, $HCO_2H$, for which

$$K_a = 1.58 \times 10^{-4}\,mol\,dm^{-3}$$

(b) Calculate the hydrogen ion concentration in a solution of sodium hydroxide which has a pH of 11. What is the concentration of the NaOH?

$$K_w = 1 \times 10^{-14}\,mol\,dm^{-3}$$

5 (a) What is meant by the term *buffer solution*?

(b) Calculate the pH of a buffer solution which contains the weak, monoprotic acid, propanoic acid ($CH_3CH_2CO_2H$), in concentration $0.1\,mol\,dm^{-3}$, and sodium propanoate concentration $0.05\,mol\,dm^{-3}$. The $K_a$ value of propanoic acid is

$$1.26 \times 10^{-5}\,mol\,dm^{-3}$$

(c) Explain how the solution in *(b)* fulfils its buffer function.

6 (a) Methanoic acid, $HCO_2H$, has a $K_a$ value of $1.58 \times 10^{-4}\,mol\,dm^{-3}$. What ratio of concentrations of methanoic acid and sodium methanoate would give a buffer of pH 4?

(b) Given the following $K_a$ values, how would you prepare a buffer solution of pH 4.44?

| Acid | $K_a/mol\,dm^{-3}$ |
| --- | --- |
| methanoic $HCO_2H$ | $1.58 \times 10^{-4}$ |
| ethanoic $CH_3CO_2H$ | $1.80 \times 10^{-5}$ |
| propanoic $CH_3CH_2CO_2H$ | $1.26 \times 10^{-5}$ |

7   (a) Sketch curves to show the change in pH as $25.0 \, cm^3$ of $0.100 \, mol \, dm^{-3}$ solutions of :

   (i)  hydrochloric acid;

   (ii) ethanoic acid  $(K_a = 1.8 \times 10^{-5} \, mol \, dm^{-3})$

   are titrated with $0.100 \, mol \, dm^{-3}$ solution of sodium hydroxide.

   (b) Which of the following indicators would be most suitable for the titration in (a)(ii)? Give reasons for your choice.

| Indicator | $K_a/mol \, dm^{-3}$ |
|---|---|
| bromothymol blue | $3.2 \times 10^{-7}$ |
| methyl red | $7.9 \times 10^{-6}$ |
| phenolphthalein | $5.0 \times 10^{-10}$ |

# Organic chemistry – general principles

**Organic chemistry** is the study of the chemistry of the compounds of carbon.

## What is organic chemistry?

The element carbon is able to form a vast number of compounds when in combination with hydrogen alone and still more when other elements, such as oxygen, nitrogen and halogens, are introduced into the molecules.

Carbon has the ability to **catenate** (make chains), that is, the ability to form covalent bonds with itself which are particularly stable as shown by the bond energies in Table 5.1. Hydrocarbon chains form the basis of most organic molecules. Stable molecules containing long chains of carbon atoms can form, can as certain ring structures. Carbon is unique in the extent to which catenation can occur. Organic compounds are the main constituents of all animal and plant life but they are **thermodynamically unstable** in the presence of oxygen and combustion reactions are exothermic, for example:

$$CH_4(g) + 2O_2(g) \rightarrow CO_2(g) + 2H_2O(g) \quad \Delta H^\theta = -882 \, kJ \, mol^{-1}$$

**Table 5.1** *Some bond energies compared to the C–C bond*

| Bond | Bond energy/kJ mol$^{-1}$ |
|------|---------------------------|
| C–C | 347 |
| N–N | 158 |
| O–O | 144 |
| Si–Si | 226 |

Life on Earth therefore exists despite a very 'hostile' environment and would appear to be in imminent danger of spontaneous combustion. Fortunately, the activation energies of the reactions with oxygen are very high and so organic compounds are **kinetically stable** at the sort of temperatures encountered on Earth.

A few compounds of carbon such as the oxides, carbonates and chlorides are not usually studied as organic chemistry and will be considered in the appropriate sections elsewhere.

## Homologous Series

The study of such a vast number of compounds would be well nigh impossible were it not for the fact that the compounds can be arranged into a relatively small number of groups or families known as **homologous series**. These have certain features in common:
- there is a general formula for the series
- all members have similar chemical properties
- there is a gradual variation (a gradation) in physical properties.

The reactions of one simple member of a series can be considered to be representative of the reactions of all members of the series. The amount of learning necessary is therefore considerably reduced. A variety of homologous series will be studied in this module and more in Module 4.

## Types of formulae

Three main types of formulae will be encountered in the study of organic compounds.

- **Empirical formula** – the formula which shows the simplest whole number ratio of the atoms present in one molecule.
- **Molecular formula** – the formula which shows the actual number of each atom in one molecule.
- **Structural formula** – the formula which shows how the various atoms are bonded to each other within the molecule.

The **empirical formula** is only of use since it can be calculated from a knowledge of the percentage composition by weight of the compound, which in turn can be found from an analysis of the compound.

For example, for a compound which contains 82.75% carbon and 17.25% hydrogen by mass, the empirical formula can be calculated to be $C_2H_5$. The method for doing this was shown in Module 1.

The molecular formula is a whole number multiple of the empirical formula. Thus it must be $(C_2H_5)_n$, where $n$ is a positive integer, and the relative molecular mass of the molecule must be $[(2 \times 12) + (5 \times 1)] \times n = 29n$. The actual relative molecular mass of the compound must then be found experimentally. This can be done in a number of ways, including mass spectrometry using the mass of the molecular ion. In the example above the relative molecular mass is found to be 58. Hence $29n = 58$ and the value of $n$ is 2. The molecular formula is therefore $C_4H_{10}$.

Determination of the structural formula requires more information either from mass spectrometry or from a knowledge of the reactions of the compound.

## Bonding in organic compounds.

The bonding in organic compounds is almost always covalent, that is, one pair of shared electrons between carbon atoms. Double and even triple covalent bonds are also commonly found. A dot and cross representation of the bonding in some simple molecules are:

An **empirical formula** shows the simplest whole number ratio of the atoms present in one molecule

A **molecular formula** shows the actual number of atoms present in one molecule.

methane                ethane

The covalent bond is more usually represented by a single line, thus:

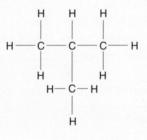

Carbon atoms always show a valency of four in organic molecules. As a result one possible structural formula for the molecular formula $C_4H_{10}$ is:

The method of representation is often described as 'showing all covalent bonds'. A more condensed way of writing the formula that still shows the groupings around each carbon atom, and hence its structure, is $CH_3CH_2CH_2CH_3$.

## Structural isomerism

**Structural isomerism** occurs when two or more different *structural* formulae can be written for the same *molecular* formula.

For the molecule $C_4H_{10}$ another perfectly plausible structure could be:

**Structural isomerism**
occurs when two or more
different structural
formulae can be written
for the same molecular
formula.

The condensed version would be written as $CH_3CH(CH_3)_2$. These two structures are said to be structural isomers. There are several ways in which this structural isomerism can occur and these will be dealt with in the relevant places in the text. Each structure must have a different name in order to identify it. With such a vast number of possible structures some systematic way of doing this is required and this will be introduced gradually as it applies to the different homologous series studied in this module.

## An orbital view of bonding

A more advanced theory of covalent bonding than that given earlier is based on the atomic orbital approach. A covalent bond is considered to be formed by the overlap of two atomic orbitals, each containing a single electron which must be of opposite spin. This results in the formation of a molecular orbital containing the shared pair of electrons which is a single covalent bond. The greater the degree of overlap between the two atomic orbitals, the stronger is the covalent bond which is formed.

The electronic structure of a carbon atom in its ground state is $1s^2 2s^2 2p^2$ and there are only two unpaired electrons available for bonding. If one electron is promoted from the 2s orbital to the 2p orbital, four unpaired electrons become available (Figure 5.1). The energy required to do this is more than compensated for by the energy released in the formation of four bonds instead of two.

These electrons are not all equivalent since they are in different types of orbital. Thus when they each overlap with the 1s atomic orbital of a hydrogen atom, as in methane, three of the bonds would be of different length to the other one.

## The formation of single covalent bonds

Consider the simple molecule of methane $CH_4$ which has four single bonds symmetrically arranged. The bonding can be understood if the excited carbon atom is considered to have its four orbitals rearranged into four new orbitals which are all equivalent. This process is referred to as **hybridisation** and the resulting orbitals as **sp$^3$ hybrid orbitals**. The shape of the hybrid orbitals is similar to a p atomic orbital except that one lobe is bigger than the other. It is the larger lobe which overlaps linearly with the 1s atomic orbital of a hydrogen atom forming a so called σ (or sigma) bond as shown in Figure 5.2.

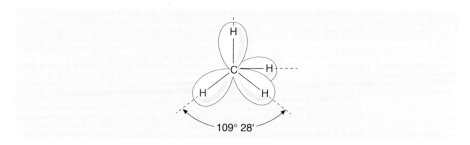

sp$^3$ hybrid orbital          s orbital of hydrogen          sigma C–H bond

*Fig. 5.2 The formation of a sigma C–H bond*

Thus all C–H bonds are equivalent and contain an electron pair. These repel each other to give a regular tetrahedral shape to the methane molecule, as shown in Figure 5.3.

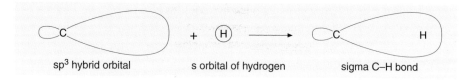

*Fig. 5.3 The tetrahedral shape of the methane molecule*

This type of hybridisation occurs in all compounds where carbon atoms form four single covalent bonds and the arrangement of the bonds around each carbon atom is tetrahedral.

## The formation of double covalent bonds

The carbon atoms which take part in double bond formation, as in ethene $C_2H_4$, undergo a different type of hybridisation. In this case the 2s and *two* of the 2p orbitals of the excited carbon atom are hybridised to form *three* new hybrid orbitals known as **sp$^2$ hybrid** orbitals which are the same shape as the sp$^3$ hybrid orbitals and lie in one plane at angles of 120°. This leaves one p orbital on the carbon atom,

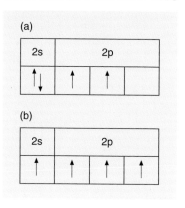

*Fig. 5.1 Electronic arrangements in the second energy level for carbon atoms (a) in the ground state and (b) in the excited state*

which lies at right angles to the plane of the $sp^2$ hybrid orbitals and which contains a single unpaired electron. When two such atoms combine to form the molecule of ethene, linear overlap between an $sp^2$ hybrid orbital from each carbon atom leads to the formation of a sigma bond between the two carbon atoms.

The other two $sp^2$ hybrid orbitals on each carbon atom can then undergo linear overlap with the 1s orbitals of four hydrogen atoms giving rise to four sigma bonds between the carbon and hydrogen atoms. The basic $C_2H_4$ molecule is thus formed and all the sigma bonds are in the same plane and at angles of 120° to each other, as shown in Figure 5.4.

The remaining p orbitals are at 90° to this plane and the positions are such that they are able to undergo lateral overlap leading to a different kind of covalent bond known as a π (or pi) bond (see Figure 5.5). This comprises two 'sausage-shaped' regions of negative charge, one above and one below the plane of the sigma bonds. This therefore explains the formation of the second bond between the carbon atoms which is completely different from the first.

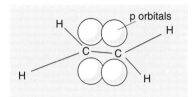

*Fig. 5.4 Two carbon atoms linked by overlap of $sp^2$ hybrid orbitals (shown as lines). The p orbitals are at right angles to the plane of the hybrid orbitals.*

## The mechanism of a reaction

A mechanism is an attempt to explain the actual steps by which a chemical reaction takes place. It attempts to show which bonds break and how they break as well as which new bonds are formed. The mechanism for the reaction is only a theory but it must fit the available experimental evidence, much of which comes from a study of the kinetics of the reaction. A variety of different mechanisms will be discussed in detail in the ensuing chapters but it would be appropriate to introduce some of the terminology and symbolism used at this point.

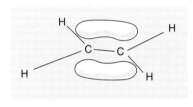

*Fig. 5.5 The pi bond formed by lateral overlap of the non-hybridised p atomic orbitals of carbon*

### Breaking covalent bonds

It is an essential part of any reaction that some of the existing covalent bonds have to be broken. The breaking of a covalent bond is referred to as **fission** and there are two ways in which this can happen:

- **heterolytic fission** – the bond is broken so that one atom of the bond gains both electrons from the bond, resulting in the formation of ions.
- **homolytic fission** – the bond is broken so that each atom of the bond has one of the electrons, resulting in **free radicals**.

A positive ion is formed when an atom loses the half share in an electron pair which had formed the covalent bond. A negative ion is formed when an atom gains both electrons which were previously shared with another atom.

- **Carbocations** or **carbonium ions** are ions where the positive charge is sited on a carbon atom.
- **Carbanions** are species where the negative charge is sited on the carbon atom.
- **Free radicals** are species which have a single unpaired electron.

Heterolytic fission is represented by a curly arrow ⌒, positioned so as to start where the electron pair is *before* the bond breaks and point to where the electron pair is *after* the bond is broken. Homolytic fission is similarly represented by a 'half arrow' ⌒.

For example,

$$H_3C - CH_3 \longrightarrow H_3C^+ + {}^-CH_3$$

represents heterolytic fission of C–C bond to give a carbocation and a carbanion.

$$H_3C - CH_3 \longrightarrow H_3C\cdot + \cdot CH_3$$

represents homolytic fission to give two free radicals.

### Types of reagent

Most types of reagent are in the main going to form new covalent bonds. Hence there are two main types of reagent:

- **nucleophiles** – species which seek out positive centres and must have a lone pair of electrons which they can donate to form a new covalent bond.
- **electrophiles** –species which seek out negative centres and must be capable of accepting a lone pair of electrons to form a new covalent bond.

## Questions

1   (a)  What is meant by the term *homologous series*?

    (b)  Look up five different homologous series and for each give:

        (i)  the general formula;

        (ii)  the formulae and names of the first five members of the series.

2   (a)  What is meant by the term *structural isomerism*?

    (b)  Deduce the structural formulae of all isomers of $C_5H_{12}$.

3   (a)  Define the terms:

        (i)  homolytic fission;        (iv)  nucleophile;

        (ii)  heterolytic fission;        (v)  free radical;

        (iii)  electrophile;        (vi)  carbocation.

    (b)  Write out as fully as you can, what is meant by the following equations:

$$H_3C - CH_2CH_2CH_3 \rightarrow H_3C^+ + {}^-CH_2CH_2CH_3$$

$$H_3C - CH_2CH_3 \rightarrow H_3C\cdot + \cdot CH_2CH_3$$

4   Describe in orbital terms how carbon atoms can form:

    (a)  single covalent bonds with each other;

    (b)  double covalent bonds with each other.

5   (a)  What is meant by the terms *empirical formula* and *molecular formula*?

    (b)  A compound contains 52.2% carbon, 13% hydrogen and 34.8% oxygen by weight and has a relative molecular mass of 46. Calculate its empirical and molecular formulae.

    Attempt to write two different structures for this molecule.

**Nucleophiles** are species which seek out positive centres and must have a lone pair of electrons which they can donate to form a new covalent bond.

**Electrophiles** are species which seek out negative centres and must be capable of accepting a lone pair of electrons to form a new covalent bond.

# 6

# Alkanes and alkenes

## Alkanes

The compounds known as **alkanes** form what is generally regarded as the simplest homologous series. They must not be underestimated however, since they are the source of very many organic chemicals and are of great commercial and economic importance. They also form the basis of the systematic nomenclature for all other aliphatic organic substances.

## General formula

$$C_nH_{2n+2}$$

## Members and nomenclature

The formulae of alkanes are simply obtained by inserting ascending integral values for $n$, beginning with 1. The first ten members are shown in Table 6.1.

**Table 6.1** *The first ten alkanes*

| $n$ | Formula | Name |
|-----|---------|------|
| 1 | $CH_4$ | methane |
| 2 | $C_2H_6$ | ethane |
| 3 | $C_3H_8$ | propane |
| 4 | $C_4H_{10}$ | butane |
| 5 | $C_5H_{12}$ | pentane |
| 6 | $C_6H_{14}$ | hexane |
| 7 | $C_7H_{16}$ | heptane |
| 8 | $C_8H_{18}$ | octane |
| 9 | $C_9H_{20}$ | nonane |
| 10 | $C_{10}H_{22}$ | decane |

The names all follow the general name for the series and end with the suffix **-ane**. The prefix **alk-** changes to indicate the number of carbon atoms in the molecule. From pentane onwards the prefix is obviously derived from the Greek, but the first four retain the names originally given to them. In the systematic nomenclature **meth-, eth-, prop-** and **but-**, will always refer to chains of one, two, three and four carbon atoms, respectively.

The molecular formula shows an increase of $-CH_2$ from one member to the next, and this is referred to as the **homologous increment**. As a result, the relative molecular masses increase by units of 14 as the series is ascended.

It would be appropriate to mention here the names of certain so-called alkyl groups which will be encountered when dealing with structural isomerism. The general formula for these is $-C_nH_{2n+1}$. They are not capable of independent existence but they do occur within other molecules. They are named in the same way as the corresponding alkane except that the name ends in **-yl** instead of -ane. The two main ones are:

- $-CH_3$, the **methyl** group
- $-C_2H_5$ or $-CH_2CH_3$, the **ethyl** group.

## Physical properties

The physical properties of the alkanes show a gradual variation as the homologous series is ascended. This does not mean, however, that the increase is necessarily linear. Table 6.2 shows the boiling points and melting points for the first eight members. A graph of boiling point against the number of carbon atoms in the molecule shows a smooth increase. This is not true for a similar graph of melting point against number of carbon atoms.

**Table 6.2** *Boiling and melting points of the first eight alkanes*

| Name | Boiling point/°C | Melting point/°C |
| --- | --- | --- |
| methane | −162 | −182 |
| ethane | −89 | −183 |
| propane | −42 | −188 |
| butane | −0.5 | −138 |
| pentane | 36 | −130 |
| hexane | 69 | −95 |
| heptane | 98 | −91 |
| octane | 126 | −57 |

*Fig 6.1 Alkanes can occur naturally. North Sea natural gas is largely methane.*

It is obvious from Table 6.2 that the alkanes gradually change their physical state at room temperature as relative molecular mass increases. Thus $C_1$ to $C_4$ are colourless gases, $C_5$ to $C_{15}$ are colourless liquids and above $C_{15}$ they are white waxy solids. They are very common materials in everyday life, for example, propane gas and butane gas are used as mobile sources of heat and light for camping, etc. They are actually sold in the liquid state under pressure in cylinders. Methane is piped into almost every home, directly from the North Sea, for cooking and heating purposes. The liquid alkanes are used in petrol for cars, paraffin, etc.

*Fig 6.2 Typical products containing alkanes*

The majority of the uses outlined in the previous paragraph require the combustion of the alkane in order to release the energy from the molecules. It is important to realise, however, that alkanes are the main source of many organic materials, including many plastics, synthetic fibres, detergents, etc. Quite a few inorganic materials, such as ammonia and sulphuric acid, are also manufactured from natural gas.

## Sources of alkanes

The main source of alkanes is from underground deposits of crude oil or natural gas, both of which require drilling in order to extract the raw materials. Sources are currently plentiful, but are thought to be finite.

Natural gas requires little purification before being used, but crude oil is a very complex mixture of alkanes and other hydrocarbons – which require separation into the pure components. This is achieved initially by fractional distillation of the crude oil, the principles of which are dealt with in Module 4, *Organic Pathways*.

## Bonding in alkanes

The carbon atoms in alkanes are $sp^3$ hybridised and are joined to each other, and to the hydrogen atoms, by sigma bonds. Since each sigma bond contains a pair of electrons, the mutual repulsion results in a tetrahedral arrangement. Structural formulae often show the bonds in such a way that the molecule appears to be planar and the molecules are often referred to as 'straight chain' molecules, for example:

Where it is necessary or desirable to show the three-dimensional nature of the bonding, a dashed line (- - -) will be used to indicate a bond going *into* the page and a wedge-shape (◁) to indicate a bond *coming out* of the page. Ordinary lines will indicate bonds *within* the plane of the page. Thus the two structures above could be shown as follows:

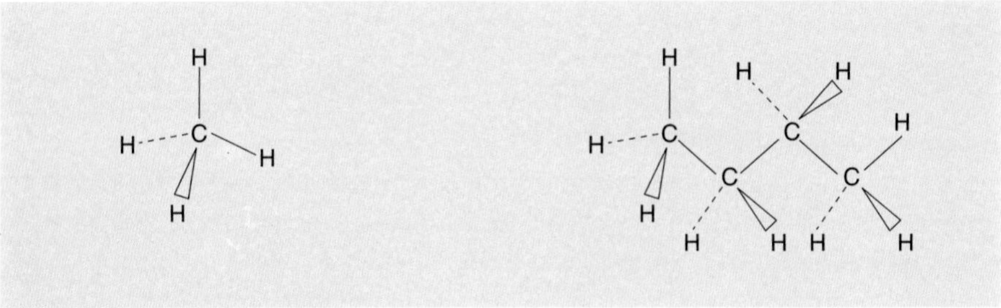

## Structural isomerism

In alkanes, the only way in which different structures can be obtained is by rearranging the carbon chain. The structure in which all the carbon atoms are joined together in a continuous chain is known as the 'straight chain' isomer (although the molecule is not actually linear as explained in the previous section). Since the structural formula does not normally represent the exact shape of the molecule, bending the chain does not actually change the length of the carbon chain.

Other isomers are formed when some carbon atoms are not part of the main chain but form '**side chains**' of different lengths. These are referred to as **branched chain** molecules. The side chains can never be of greater length than the main chain.

The first alkane to show isomerism is $C_4H_{10}$, which can have two structures:

$$CH_3CH_2CH_2CH_3 \quad \text{and} \quad \overset{\displaystyle CH_3}{\underset{\displaystyle CH_3CHCH_3}{|}} \quad \text{(or } CH_3CH(CH_3)_2)$$

butane        2-methylpropane

The **systematic nomenclature** for organic compounds uses a name based on the longest continuous carbon chain. Alkyl groups are regarded as substituents and their names come before the name of the longest carbon chain and they are preceded by a number to indicate the carbon atom at which the substitution has occurred. The number does not refer to the number of alkyl groups being substituted.

Thus the name 2-methyl propane indicates that a methyl group has been substituted on the second carbon atom of a three carbon chain. The prefix 2- is not strictly necessary in this case since the second carbon atom is the only atom where substitution by a methyl group could occur without increasing the carbon chain to four carbon atoms again.
The next alkane $C_5H_{12}$ has three isomers:

$$CH_3CH_2CH_2CH_2CH_3 \qquad \overset{\displaystyle CH_3}{\underset{\displaystyle CH_3CH_2CHCH_3}{|}} \qquad \overset{\displaystyle CH_3}{\underset{\displaystyle \underset{\displaystyle CH_3}{\overset{\displaystyle |}{CH_3CCH_3}}}{|}}$$

$$CH_3(CH_2)_3CH_3 \qquad\qquad CH_3CH_2CH(CH_3)_2 \qquad\qquad CH_3C(CH_3)_3$$

pentane            2-methylbutane        2,2-dimethylpropane
(b.p. 36 °C)        (b.p. 28 °C)           (b.p. 10 °C)

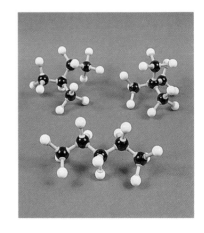

*Fig 6.3 The three structural isomers of $C_5H_{12}$*

The main carbon chain can be numbered from either end and should be done so as to produce the lowest number, for the substituents (2-methylbutane rather than 3-methylbutane in the example above). Where there is more than one substituent, each one must be given a number. Any name must be unambiguous. It would be good practice to make models of these different structures to see how the overall shape varies.

All isomers of alkanes have similar chemical properties but differ in physical properties such as boiling point, and it is this that allows them to be isolated in the pure state. The boiling point is determined by the strength of the intermolecular forces of attraction in the liquid state. The only intermolecular forces present in alkanes are van der Waals forces and the strength of these depends on the number of electrons exposed on the outside of the molecule. The number is greatest in straight chain molecules and decreases as the degree of branching increases. Hence boiling points decrease as degree of branching increases, as shown in the examples above.

### Cyclic alkanes

A number of alkane molecules can be formed where the ends of the carbon chain have been linked up to form a 'ring' or 'cyclic' structure, for example:

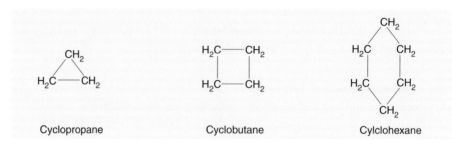

Cyclopropane          Cyclobutane          Cylclohexane

The bond angles for the smaller rings are very different from the 109° 28' of the $sp^3$ hybrid orbitals, which indicates poor overlap of the orbitals and considerable strain in the ring. They behave as normal alkanes as far as the C–H bond is concerned, but would be much more reactive towards reagents which break the C–C bond.

## Reactions of alkanes

### Combustion

All alkanes, and indeed all hydrocarbons, will burn in air or oxygen with the release of heat energy and many are used for this purpose. The products of the complete combustion in excess oxygen are always carbon dioxide and water. As a result equations can be simply constructed, for example, the equation for combustion of methane is:

$$CH_4(g) + 2O_2(g) \rightarrow CO_2(g) + 2H_2O(g) \qquad \Delta H = -882 \, kJ \, mol^{-1}$$

The same kind of reaction is used to heat homes, drive motor cars, power aeroplanes, and for many other purposes.

If the alkane and oxygen are mixed before ignition, an explosion will result. Great care needs to be taken to avoid this in coal mines where methane is released naturally and can build up to explosive proportions. In car engines however just such a small explosion is carried out in order to drive the engine. The advantages and disadvantages of using liquid as opposed to gaseous fuels will be considered in Module 4, *Organic Pathways*.

### Halogenation

Halogenation is the introduction of a halogen atom into the alkane molecule. It occurs by a **substitution reaction**, which means that a halogen atom replaces one or more of the hydrogen atoms in the alkane. Hydrogen halides are always produced in these reactions (as 'steamy' acidic fumes) together with a product called a **haloalkane**.

All halogens react with all alkanes, but the reaction is quicker with chlorine than with bromine which is in turn quicker than with iodine. Chlorine reacts explosively but iodine results in an equilibrium. The rate of reaction also decreases as the relative molecular mass of the alkane increases. The presence of sunlight, ultraviolet light or some other form of energy is essential for these reactions to proceed at a reasonable rate. This can easily be

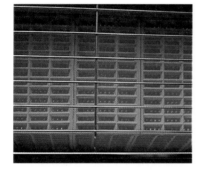

*Fig 6.4 Domestic gas fires make full use of the combustion of alkanes.*

A **substitution reaction** is one in which an atom or group of atoms in one molecule is replaced by another atom or group of atoms.

demonstrated experimentally using hexane and bromine and placing the same mixture in two test tubes, one of which is wrapped in black paper. Exposure to sunlight or a strong artificial light leads to a rapid decolorisation of the bromine in the uncovered test tube while the colour remains for much longer in the darkened one. Examples of such reactions are:

$$CH_4 \;+\; Cl_2 \;\rightarrow\; CH_3Cl \;+\; HCl$$
$$\text{methane} \qquad\qquad \text{chloromethane} \quad \text{hydrogen chloride}$$

$$C_6H_{14} \;+\; Br_2 \;\rightarrow\; C_6H_{13}Br \;+\; HBr$$
$$\text{hexane} \qquad\qquad \text{bromohexane} \quad \text{hydrogen bromide}$$

There is no way of determining which hydrogen atom will be replaced since all the C–H bonds are equivalent. It is not possible therefore to make a specific haloalkane by this method since it is impossible to stop further substitutions taking place.

## Mechanism for the chlorination of methane

The mechanism for the chlorination of methane is **homolytic free radical substitution**. The proposed mechanism for this reaction is as shown below:

the initiation step

the propagation steps

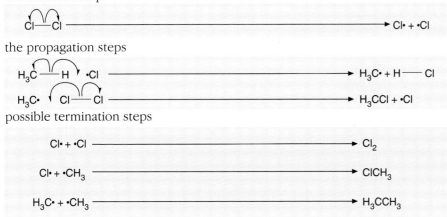

possible termination steps

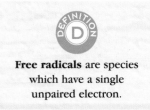

**Free radicals** are species which have a single unpaired electron.

The **initiation step** is the reaction which first generates the free radicals. Energy is required to break this bond, which is why UV light is necessary.

The **propagation steps** are reactions which are instigated by free radicals but also regenerate *more* free radicals, so allowing the reaction to continue unaided.

The **termination steps** are reactions in which free radicals are used up and not regenerated. As a result, the reaction stops eventually.

Free radicals are extremely unstable and reactive species. The reaction between chlorine and methane occurs explosively unless the amount of UV radiation is controlled by using subdued sunlight.

A variety of other products, such as chloromethanes and chloroethanes, are produced, in very small quantities, by a continuation of the same mechanism.

## Alkenes

This is another homologous series containing carbon and hydrogen only.

### General formula

$$C_nH_{2n}$$

Each molecule contains two hydrogen atoms less than the corresponding alkane.

### Members and nomenclature

The names of the alkenes follow exactly from the general name for the series. Thus, compared to alkanes, the suffix simply changes to **-ene** whilst the prefix remains the same, indicating the number of carbon atoms. There is no alkene corresponding to $n = 1$, so the first two members of the series are:

ethene     $C_2H_4$   or   $H_2C = CH_2$
propene     $C_3H_6$   or   $CH_3CH = CH_2$

All alkene molecules contain one carbon–carbon double bond at some point in the carbon chain and as a result are said to be **unsaturated**.

### Structural isomerism.

Structural isomerism can occur by moving the double bond to different positions in the carbon chain. The position of the double bond is then indicated by a number inserted between the prefix and the -ene, for example $C_4H_8$ could be

$CH_3CH_2CH = CH_2$    or    $CH_3CH = CHCH_3$
but-1-ene              but-2-ene

Branching of the carbon chain is still possible as the length of the carbon chain increases. In these cases, the names are still based on the longest continuous carbon chain present, as shown by *some* of the isomers of $C_5H_{10}$ below:

$CH_3CH_2CH_2CH = CH_2$    $(CH_3)_2CHCH = CH_2$    $(CH_3)_2C = CHCH_3$
pent-1-ene          3-methylbut-1-ene      2-methylbut-2-ene

Note that the position of the double bond takes precedence in numbering the carbon atoms of the longest carbon chain. As a result, the isomer 3-methylbut-1-ene is not called 2-methylbut-3-ene.

Such structural isomers differ from each other only in physical properties; their chemical reactions are all the same. Apart from this structural isomerism, a new kind of isomerism (geometric isomerism) occurs in alkenes, which is discussed later.

### Bonding in alkenes

The two carbon atoms of the double bond are sp$^2$ hybridised and form a double covalent bond with each other, one bond of which is a sigma bond and the other a pi bond. The arrangement of the bonds around these two carbon atoms is planar with bond angles of 120°.

The pi bond between the carbon atoms is therefore completely different from the sigma bond and consists of two regions of negative charge, one above and one below the plane of the carbon and hydrogen atoms.

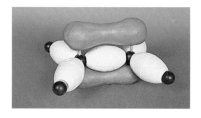

*Fig 6.5 An orbital model of ethene*

Some experimental evidence to indicate that the two bonds in a carbon–carbon double bond are not the same includes the following.
- The bond energy of C=C ($612 \, \text{kJ mol}^{-1}$) is greater than C–C ($348 \, \text{kJ mol}^{-1}$) but not twice as big. Hence the pi bond is weaker than the sigma bond.
- The greater strength of the C=C bond is supported by the shorter bond length (due to the greater overlap) which is 0.134 nm as opposed to 0.154 nm for C–C.
- The existence of geometric isomers, as explained below.

Carbon atoms in the alkene molecule other than the two directly engaged in the formation of the double bond will probably be sp$^3$ hybridised and the arrangement of the bonds around these carbon atoms will be tetrahedral as usual.

## Geometric isomerism

**Geometric isomerism** is a different kind of isomerism which exists as a direct consequence of the double bond. A single sigma bond between carbon atoms will allow rotation around the axis of the bond without any reduction in the degree of overlap. As a result, there is free rotation about this bond since no covalent bonds need to be broken. With a double bond however, such rotation would lead to a decrease in the degree of overlap of the pi bond and consequently cannot be achieved without the supply of a suitable quantity of energy. There is restricted rotation about the carbon–carbon double bond.

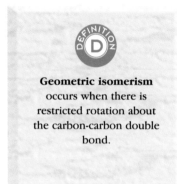

**Geometric isomerism occurs when there is restricted rotation about the carbon-carbon double bond.**

The isomerism depends on the arrangement of the groups or atoms around the double bond, for example, but-2-ene can have two structures:

cis–but–2–ene          trans–but–2–ene

The prefixes *cis-* and *trans-* are used to distinguish between these isomers. *Cis-* refers to two groups which are the same being on the same side of the double bond, while *trans-* refers to them being on opposite sides of the double bond. These isomers would not normally be regarded as structural isomers since they have the same basic carbon chain with the atoms linked in the same order. It is only the orientation of the groups in space around the double bond which is different.

The general term for isomerism relying on the orientation of groups is **stereoisomerism** and geometric isomerism is one form of this. Such isomers have only very slight differences in physical properties such as boiling points.

This kind of isomerism is to be found in any molecule of the type

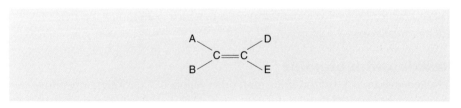

where A is not the same atom or group as B and D is not the same as E. Some further examples are:

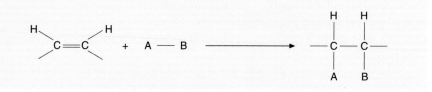

cis–1,2–dichloroethene                    trans–1,2–dichloroethene

There is also a structural isomer, 1,1-dichloroethene.

cis–3–methylpent–2–ene                    trans–3–methylpent–2–ene

There will also be many structural isomers.

# Reactions of alkenes

As shown earlier, the bonding in alkenes is quite different from that in alkanes and it is therefore not surprising that their reactions are also quite different. The pi bond in alkenes is much more available to attack and is also a weaker bond than the sigma bonds in alkanes. Hence the alkenes will react with a greater variety of reagents and will react much more readily than alkanes.

## Addition reactions

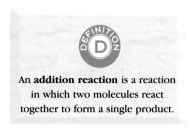

An **addition reaction** is a reaction in which two molecules react together to form a single product.

Addition reactions are the most typical of the alkene reactions although they are not the only kind of reactions which alkenes undergo. An addition reaction is one in which two molecules react together to form a single product.

In the addition reactions of alkenes, it is always the pi bond that breaks in order to release electrons to form bonds with the reactant molecule. The sigma bonds never break so that the two carbon atoms of the double bond always remain bonded together and a saturated molecule is formed from an unsaturated alkene. The equations can therefore always be written as follows:

These reactions occur at room temperature, which is an indication that they are easy reactions to perform.

## Reaction with bromine

All alkenes will react with bromine (and other halogens) at room temperature. Bromine is not usually used in the pure state since it presents a safety hazard, but the hazard can be avoided by dissolving it in an organic solvent such as hexane.

Typical reactions are as follows:

$$H_2C = CH_2 + Br_2 \rightarrow CH_2BrCH_2Br$$
ethene $\qquad\qquad$ 1,2-dibromoethane

$$CH_3CH = CHCH_3 + Br_2 \rightarrow CH_3CHBrCHBrCH_3$$
but-2-ene $\qquad\qquad$ 2,3-dibromobutane

The products of the reactions are named as if they were substitution products of an alkane, even though the reaction performed is addition to an alkene. The name is always based on the structure of the molecule, using the longest carbon chain present, not on the reaction by which it is formed.

## Test for unsaturation

If a reaction is to be used as a test it should, generally speaking, be capable of producing an observable result in a reasonable time and in a relatively simple piece of apparatus.

Since the products of all these reactions are colourless, a test for unsaturation is the decolorisation of brown bromine solution.

## Reaction with hydrogen bromide

The reaction here is exactly as predicted from the general equation. Hence the reaction with ethene at room temperature is as follows:

$$H_2C = CH_2 + HBr \rightarrow CH_3CH_2Br$$
bromoethane
(a colourless liquid)

## Mechanism of addition reactions

Both of the reactions given and all the other addition reactions for alkenes occur by a mechanism known as **heterolytic electophilic addition**. The mechanism for the reaction of ethene with hydrogen bromide is as follows:

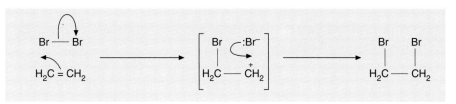

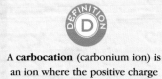

A **carbocation** (carbonium ion) is an ion where the positive charge is sited on the carbon atom.

The electrophile here is the $H^{\delta+}$–$Br^{\delta-}$ molecule which is already polarised and so is attracted to the negative region which is the pi bond. In the case of bromine polarisation must be induced in the molecule by some means before a reaction will occur.

The mechanism for the reaction of ethene with bromine is as follows:

The carbocation is not a species which can be isolated. It is very unstable and is simply a transition state through which the reaction passes. It does, however, represent the activation energy peak for the reaction and hence it is specified, but only in brackets to indicate its transitory nature. Note that the Br⁻ ion acts as a nucleophile in the second part of the reaction. This cannot occur until the carbonium ion has been formed by electrophilic attack, hence the designation of this mechanism as electrophilic addition.

## Poly(ethene)

Poly(ethene) is a **polymer**, that is, a very large molecule and, since it is made from a single unit or **monomer** by a process of **addition**, it is known as an **addition polymer**.

The single monomer unit is ethene, molecules of which are made to join together (by a free radical mechanism) to form very long molecules which are essentially alkanes. The basic reaction is:

$$n\mathrm{CH_2} = \mathrm{CH_2} \rightarrow (-\mathrm{CH_2}-\mathrm{CH_2}-)_n$$

Poly(ethene), more commonly known as 'polythene', is a plastic material which is in everyday use for many different purposes such as film packaging, electrical insulation, containers for household chemicals such as washing-up liquid, buckets, food boxes, washing-up bowls, etc. This is obviously a very important industrial application of alkenes.

In fact, there are different forms of poly(ethene) the two main ones being: low density poly(ethene) and high-density poly(ethene).

Low-density poly(ethene) is made by subjecting the ethene to a very high pressure (1000–3000 atm) at a moderate temperature in the range 420–570 K. The average polymer molecule contains between $4 \times 10^3$ and $40 \times 10^3$ carbon atoms, with many short carbon chain branches.

High density poly(ethene) is made by subjecting ethene to a lower temperature (310–360 K) and pressure (1–50 atm) in a suspension of titanium(III) or titanium(IV) chloride and an alkylaluminium compound such as triethyl aluminium (known as a Ziegler–Natta catalyst). This type of poly(ethene) has few branched chains and the molecules can pack more closely together. It is therefore more dense and melts at a higher temperature. It has a crystalline structure which gives it a rigid structure of considerable strength.

## The problem of disposal

Polythene, along with many other polymers, provides us with many very useful materials. It does, however, present a very considerable environmental problem when we come to dispose of it since it is not biodegradeable, that is, it is not broken down by the action of microorganisms. As a result, it accumulates in vast quantities in rubbish tips and will never disappear. This is a problem which needs to be solved in the near future if we are not to leave an ever increasing problem for future generations.

*Fig 6.6 Low-density polythene is widely used for packaging.*

*Fig 6.7 High-density polythene has a structure of considerable strength*

*Fig 6.8 Polythene is not biodegradable, and when burned gives off toxic fumes. This causes problems of disposal with long term effects*

# Questions

1    *(a)*   Write the structural formulae for the following:

       (i)   2-methylpropane;    (ii)   2.2-dimethylpropane;

       (iii)   2.2.3-trimethylbutane;

   *(b)*   Give the systematic names for the following:

       (i)   $CH_3CH_2CH_2CH_3$;    (ii)   $C(CH_3)_4$;    (iii)   $CH_3CH_2CH(CH_2CH_3)_2$.

2    *(a)*   Write down the structural formulae for all the isomers of $C_6H_{14}$.

   *(b)*   Write the systematic name for each isomer.

   *(c)*   Arrange the isomers as far as you can in order of increasing boiling points.

3    *(a)*   Write the mechanism for the reaction of chlorine with methane to form chloromethane.

   *(b)*   Explain briefly why small traces of dichloroethane are found in the final reaction mixture in *(a)*.

4    Use your knowledge of the mechanism in 3*(a)* to write the mechanism for the reaction of chlorine with chloromethane to form dichloromethane.

5    Give the equation, the conditions used and the names of the products when:

   *(a)*   but-2-ene reacts with hydrogen bromide;

   *(b)*   propene reacts with bromine.

   Give the mechanism for the reaction in each case.

6    *(a)*   Write down the structural formulae of all isomers of but-2-ene and name them.

   *(b)*   Explain briefly the reason for the existence of these isomers.

   *(c)*   Write down the structural formulae all the isomers of $C_4H_8$ and name them.

   *(d)*   How would you show that each of these isomers was unsaturated ?

7    Use the data from Table 6.2 on page 73 to plot graphs of

   *(a)*   boiling point against relative molecular mass

   *(b)*   melting point against relative molecular mass.

   Comment on the graphs obtained in each case.

8    *(a)*   Explain briefly why alkene molecules react with more substances than alkanes.

   *(b)*   From your knowledge of poly(ethene), deduce a possible structure for

       (i)   poly(propene),

       (ii)   poly(chloroethene) given that the formula of chloroethene is $CH_2=CHCl$.

# 7

# Haloalkanes

## Haloalkanes

Compounds formed when a member of the halogen group is substituted into an alkane are called haloalkanes.

### General formula

$$C_nH_{2n+1}X$$

X is usually chlorine, bromine or iodine.

### Members and nomenclature

The names of haloalkanes are derived from the generic name. The name of the halogen present comes first followed by the name of the alkane chain into which the halogen has been substituted. A number preceding the halogen indicates its position on the carbon chain. Some examples are:

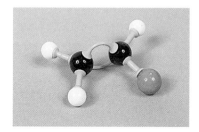

Fig 7.1 A model of the chloroethene molecule

| | |
|---|---|
| $CH_3Cl$ | chloromethane |
| $CH_3CH_2Br$ | bromoethane |
| $CH_3CH_2CH_2Br$ | 1-bromopropane |
| $CH_3CHBrCH_3$ | 2-bromopropane |

## Functional groups

When atoms or groups of atoms other than carbon or hydrogen are present in an organic molecule, they are generally much more reactive than the hydrocarbon chain, which can only react in the ways already described for alkanes. Such groups are known as **functional groups** since it is these groups which determine the reactions of the molecule. The functional group in haloalkanes is the halogen atom. The carbon–halogen bond is very much easier to break than a carbon–hydrogen bond in the hydrocarbon chain. As a result, all the reactions of haloalkanes are reactions of the halogen atom, as will be seen later.

## Isomerism

### Structural isomerism

Structural isomerism can occur in the usual ways by:
* moving the halogen atom to different positions on the carbon chain as in 1-bromopropane and 2-bromopropane above
* branching of the carbon chain in larger molecules, for example, 2-bromobutane $CH_3CH_2CHBrCH_3$ and 2-bromo-2-methylpropane $(CH_3)_2CBrCH_3$, which are both isomers of $C_4H_9Br$.

### Optical isomerism

Optical isomerism is another kind of stereoisomerism, that is, it depends on the orientation of atoms or groups relative to each other. It is not confined to haloalkanes, but it is convenient to introduce it here. Optical isomerism occurs in organic molecules containing a carbon atom which is attached to four atoms or groups of atoms that are all different from each other. Such a carbon atom is said to be **asymmetric** and the molecule is said to be **chiral**. 2-Chlorobutane would be such a molecule, the carbon atom labelled * being the asymmetric carbon atom (Figure 7.2).

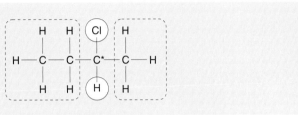

*Fig. 7.2 2-Chlorobutane contains an asymmetric carbon atom*

Two different structures of such a molecule can exist, one of which is the mirror image of the other. The two mirror images are non-superimposable. (See Figure 7.3). The two different structures are called **optical isomers** or **enantiomers** and the only difference between them is the effect they have on the plane of plane-polarised monochromatic light. One optical isomer will rotate the plane of plane-polarised light to the right and the other isomer will rotate it to the left. These are known as the (+) form and the (−) form, respectively.

An equimolar mixture of the two optical isomers would have no effect on the plane of plane-polarised light since the rotational effect of one would be cancelled out by the rotational effect of the other. Such a mixture is known as the **racemic mixture** and this is the end product of most methods of preparation. If the individual isomers are required, they must be separated from this mixture, a procedure which is quite difficult. The existence of optical isomers is of great importance in biology where one isomer will be effective in a particular process and the other will not.

## Types of haloalkane

The hydrocarbon chain to which a functional group is attached can exist in one of three possible forms. As a result, there are three different types of haloalkane, in fact, there are three different types of compound in any homologous series containing functional groups. These types are **primary**, **secondary** and **tertiary**.

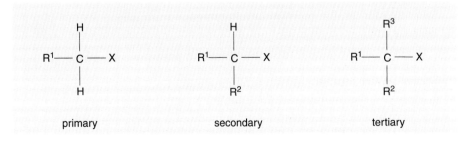

$R^1$, $R^2$ and $R^3$ are alkyl groups which may be the same or different but must contain at least one carbon atom.

A primary compound must have two H atoms directly attached to the carbon atom that carries the functional group. They are therefore characterised by having the $-CH_2X$ grouping. Secondary compounds have only one H atom attached to the functional group carbon atom and are characterised by having a   –CHX grouping. Tertiary compounds have no H atoms on the functional group carbon atom.

There is no difference in the way in which these different haloalkanes react but there is a difference in the rate at which they react as will be seen in the next section.

In other homologous series, notably alcohols, the type of compound can be of greater significance and can influence the type of reaction as well as the rate.

## Rates of reaction of haloalkanes

Although all haloalkanes react with the same reagents and undergo the same types of reaction, the rate of reaction varies with two factors.

- **The nature of the halogen**. Iodides react faster than bromides, which in turn react faster than chlorides. This is because the iodine atom is bigger than bromine and so the carbon–iodine bond is longer than the carbon–bromine bond. Hence it has a lower bond energy and is more easily broken. The same argument applies to the carbon–chlorine bond. The relevant values are shown in Table 7.1.

**Table 7.1** Relationship between bond length and bond energy

| Bond | Bond length/nm | Bond energy/kJ mol$^{-1}$ |
|---|---|---|
| C–Cl | 0.177 | 338 |
| C–Br | 0.193 | 276 |
| C–I | 0.214 | 238 |

- **The type of haloalkane**. Tertiary compounds react faster than secondary which in turn react faster than primary. This is because of the stability of the various types of carbocation as will be seen later in the discussions on mechanisms.

A tertiary iodide would be expected to react very much more quickly than a tertiary bromide and more quickly than a secondary iodide. Tertiary iodides would be the quickest of all haloalkanes to react and primary chlorides the slowest. A practical method of demonstrating this will be discussed later.

# Reactions of haloalkanes

## Nucleophilic substitution reactions

Nucleophilic substitution reactions are the most characteristic reactions of haloalkanes and involve the substitution of the halogen atom by some other atom or group. Two such reactions are discussed here.

### Reaction with sodium hydroxide solution.

- Reagent:      *aqueous* sodium hydroxide
- Conditions:  boil under reflux
- Product:      an alcohol

It is the hydroxide ions from the sodium hydroxide which attack the haloalkane. Hence, for bromoethane, the reaction is:

$$CH_3CH_2Br + OH^- \rightarrow CH_3CH_2OH + Br^-$$
$$\text{ethanol}$$

This reaction is sometimes referred to as **hydrolysis**. Note that the reagent must be in aqueous solution in contrast to the elimination reaction below.

## Reaction with potassium cyanide
- Reagent: potassium cyanide in solution in *ethanol*.
- Conditions: boil under reflux
- Product: A nitrile

The potassium cyanide is a source of cyanide ions, $CN^-$, and it is these which attack the haloalkane. Hence, for iodoethane the reaction is:

$$CH_3CH_2I + CN^- \rightarrow CH_3CH_2CN + I^-$$
$$\text{propanenitrile}$$

It is beyond the scope of this module to discuss the names or properties of nitriles but it is worthy of note that the nitrile produced contains one more carbon atom than the original haloalkane. This can be very useful in organic syntheses and will be dealt with more fully in Module 4. The mechanism for these reactions will be given later.

**Practical note**. Both reactions refer to a process known as 'boiling under reflux'. This is a technique used frequently in organic chemistry where the reactions are often slow and the reagents volatile. The technique consists of using a condenser mounted vertically on top of the reaction vessel so that any vapours escaping during the heating process will condense to a liquid and run back into the flask. As a result the heating process can be carried out for a longer period (see Figure 7.4).

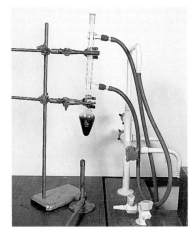

*Fig 7.3 Apparatus for heating under reflux*

## Elimination reaction
- Reagent: *Ethanolic* solution of potassium hydroxide
- Conditions: heat
- Product: an alkene

Since the product is an alkene, this reaction can only be performed on haloalkanes containing at least two carbon atoms. Hence the reaction for bromoethane is:

$$CH_3CH_2Br + OH^- \rightarrow CH_2{=}CH_2 + Br^- + H_2O$$
$$\text{ethene}$$

The yield is very small for molecules with only two or three carbon atoms.

Looking at the structure of the two organic molecules, the haloalkane appears to have lost a molecule of HBr (this is why the reaction is termed an 'elimination' reaction). It is not quite true, however, to say that a 'molecule' of HBr has been removed, since the H is removed from one carbon atom and the Br from an adjacent carbon atom.

Although the mechanism for this reaction is not required in this module, it is worth noting that in some cases it is possible to form two different alkenes in the same reaction. Consider, for example, the reaction of 2-bromobutane with 'ethanolic' potassium hydroxide. The Br atom must of course be removed from carbon atom number two, but there are two alternatives for the H atom. If it is removed from carbon atom number one, the product will be but-1-ene. If on the other hand, it is removed from carbon atom number three, the product is but-2-ene.

**Elimination reactions** are reactions in which the elements of a simple molecule such as HBr, $H_2O$, etc. are removed from the organic molecule and not replaced by any other atom or group of atoms.

The products of the reactions between a haloalkane and potassium hydroxide are quite different according to the conditions used. This shows the importance of reaction conditions in organic chemistry. In practice, both products are formed in both reactions, but the alcohol will predominate in aqueous solution and the alkene in ethanolic solution. Organic reactions rarely lead to a single pure product: purification is usually necessary and may be lengthy and costly.

## Mechanism for heterolytic nucleophilic substitution

The carbon–halogen bond differs from the carbon–carbon and carbon–hydrogen bonds present in molecules of haloalkanes in that it is polarised to a much greater extent because of the greater difference in electronegativity between the halogen atom and carbon atom. The carbon atom therefore has a partial positive charge and the halogen a partial negative charge.

The degree of polarisation will be greatest for chlorine and least for iodine since chlorine has the greater electronegativity. However, this does not determine the reactivity of the halogens, which was explained earlier.

It is this positive charge on the C atom which is attacked by nucleophiles. Both $OH^-$ and $CN^-$ have lone pairs of electrons to donate and therefore can act as nucleophiles. This attack can take place in two different ways.

### $S_N1$ mechanism
$S_N1$ means that the reaction occurs by substitution of a nucleophile and that it follows first-order kinetics. This mechanism occurs in two stages.
• Heterolytic fission of C–Br bond to form a carbocation:

The first step of this mechanism is the rate-determining step, the second step

being much faster. Hence a study of the kinetics of the reaction would be expected to show that the reaction is first order with respect to the haloalkane and zero order with respect to the OH⁻. The rate equation would then be:

$$\text{rate} = k\,[\text{haloalkane}]$$

If experimentation confirms this then the mechanism is likely to be $S_N1$.

Although no attempt has been made to show the shapes of the molecules or intermediates, it should be apparent that the initial and final molecules are tetrahedral but that the intermediate carbocation is planar in shape. This means that the intermediate carbocation can be attacked, by the nucleophile, with equal facility from either side. If the initial haloalkane molecule was chiral, for example, 2-bromobutane, equal quantities of the two chiral molecules would be formed and the product would show no optical activity.

## $S_N2$ mechanism

$S_N2$ means that the reaction occurs by substitution of a nucleophile and that it follows second-order kinetics. There is only one stage in this mechanism. The nucleophile attacks the $C^{\delta+}$ and begins to form a covalent bond with it. At the same time the C–Hal bond begins to break heterolytically. This is one continuous process with the C–nucleophile bond getting stronger and the C–hal bond getting weaker. The transition state (which corresponds to the activation energy peak) is formed when both bonds are of equal strength. In this mechanism, the formation of the transition state represents the rate-

transition state

determining step. A study of the kinetics of the reaction would be expected to show that the reaction is first order with respect to the haloalkane and first order with respect to the OH⁻. The rate equation would then be:

$$\text{rate} = k\,[\text{OH}^-]\,[\text{C}_2\text{H}_5\text{Br}]$$

## Will it be S$_N$1 or S$_N$2?

### Types of carbocation

There are three types of carbocation, primary, secondary and tertiary:

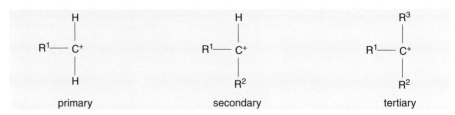

primary                    secondary                    tertiary

R$^1$, R$^2$ and R$^3$ are alkyl groups which may be the same or different. The terminology is exactly the same as that seen in the various types of haloalkane and will be seen again in other compounds.

### The stability of carbocations

Carbocations are unstable species. They can be made more stable if the charge on the cation is dispersed. Alkyl groups are more electron releasing than hydrogen atoms and so push the bonded electrons towards the C$^+$ atom, reducing its charge. The more alkyl groups there are, the more the charge will be dispersed and the greater the stability of the carbocation. Hence the order of stability of carbocations is: Tertiary more stable than secondary, secondary more stable than primary. The carbocation corresponds to the activation energy peak of the reaction. A route involving tertiary carbocations is easier, therefore, than one involving secondary or primary carbocations.

The only way to be certain which mechanism is occurring in a given reaction is to determine the orders with respect to each reagent experimentally. However, the fact that a carbocation is formed in S$_N$1 reactions means that tertiary haloalkanes are likely to react via an S$_N$1 mechanism and primary haloalkanes are likely to react via an S$_N$2 mechanism. If both mechanisms are operating, as is possible for secondary haloalkanes, the order with respect to OH$^-$ will be fractional.

## Practical tests for halogen atoms

The halogen atoms in haloalkanes are covalently bonded to a carbon atom. In order to detect their presence, they are converted to the corresponding halide ion, which is then detected in the normal way.

**Test**
1 React the haloalkane with aqueous sodium hydroxide (heat may be necessary):
$$C_nH_{2n+1}X + OH^- \rightarrow C_nH_{2n+1}OH + X^-$$
2 Acidify with dilute nitric acid to remove excess hydroxide ion
3 Add aqueous silver nitrate.

**Result**
- White precipitate (soluble in dilute ammonia)                    –Cl present
- Cream precipitate (partially soluble in dilute ammonia)          –Br present
- Yellow precipitate (insoluble in dilute ammonia)                 –I present

This test can also be done by heating the haloalkane, *dissolved in ethanol*, with silver nitrate solution. The same precipitates will occur. This method can also be used to test the reactivity of various haloalkanes by placing them in test tubes contained in a beaker of water maintained at constant temperature. The time taken for the precipitates to appear when silver nitrate solution is added indicates the rate of reaction. Sometimes the carbon–halogen bond is sufficiently weak for this reaction to occur on warming with aqueous silver nitrate solution alone.

## Questions

1   (a)  Write down the structures of all isomers of $C_4H_9Cl$

(b)  Assign a name to each isomer.

(c)  Classify each isomer as primary, secondary or tertiary.

(d)  State the type of isomerism occurring in each case and explain how it arises.

2   Write equations and give the conditions for the reaction of an ethanolic solution of potassium hydroxide with:

(a)  2-chloropropane;

(b)  2-chlorobutane.

Explain any differences in these reactions.

3   Write equations for the reactions of aqueous sodium hydroxide with:

(a)  1-chloropropane;

(b)  2-chloro-2-methylbutane;

Give the mechanism which is likely to occur in each case and give its name.

4   Describe an experiment you would perform in order to confirm that the predictions made about the mechanisms on page 85, were in fact correct.

5   A haloalkane is known to contain four carbon atoms. Describe simple experiments you would perform in order to show that:

(a)  the halogen present was bromine;

(b)  the haloalkane was secondary.

6   (a)  Explain why 2-bromobutane exhibits optical isomerism.

(b)  Reaction of the d- form of 2-bromobutane with aqueous sodium hydroxide gives a product which shows no optical activity.

Deduce and give the mechanism which must be occurring and explain the lack of optical activity by reference to the mechanism.

# 8 Functional groups containing oxygen

## Alcohols (or alkanols)

### General formula

$$C_nH_{2n+1}OH$$

This could also be written as $C_nH_{2n+2}O$ but the one above is more useful since it highlights the fact that one of the H atoms is different from the others in that it is bonded to an oxygen atom rather than a carbon atom.

### Formulae and nomenclature

The names of alcohols are simply based on the generic name of **alkanol**, as shown in Table 8.1. Isomerism occurs in the usual ways and a number is inserted before the **-ol**, to indicate the position of the –OH group on the carbon chain when this is necessary.

**Table 8.1** The names of some alcohols

| Formula | Name |
| --- | --- |
| $CH_3OH$ | methanol |
| $CH_3CH_2OH$ | ethanol |
| $CH_3CH_2CH_2OH$ | propan-1-ol |
| $CH_3CHOHCH_3$ | propan-2-ol |

### Functional group and its test

The functional group is the –OH group, the presence of which can be shown by the following test.

 To a sample of the alcohol in a clean, *dry* test tube, carefully add some solid phosphorus pentachloride ($PCl_5$). The result is 'steamy' acidic fumes of HCl are evolved.

The organic product of the reaction is always a haloalkane. The equation for the reaction in general is:

$$ROH + PCl_5 \rightarrow RCl + POCl_3 + HCl$$

and for the specific example of ethanol:

$$CH_3CH_2OH + PCl_5 \rightarrow CH_3CH_2Cl + POCl_3 + HCl$$
$$\text{chloroethane}$$

It is important that the test tube is dry since water contains an –OH group and thus produces fumes of HCl which would invalidate the test.

Note that $PCl_5$ gives HCl with any –OH compounds, e.g. ethanoic acid, $CH_3COOH$. It is not a specific test for alcohols only.

## Types of alcohol

As in previous homologous series, there are three types:
- primary      containing      $-CH_2OH$
- secondary   containing      $-CHOH$
- tertiary     containing      $-COH$

The effect of their structure on the rate at which they react is the same as discussed in Chapter 7. Unlike previous homologous series, however, the compounds can, in certain cases, behave differently towards a particular reagent (this is discussed later). All, however, give the positive test for the $-OH$ group.

# Aldehydes (alkanals) and ketones (alkanones)

These two homologous series are usually considered together since they have many reactions in common.

## General formulae

$$C_nH_{2n}O$$

Both groups have the same general formula but a more useful way of writing these shows the functional groups:

aldehydes                    ketones

$R^1$ and $R^2$ are alkyl groups which may be the same or different. In the case of aldehydes $R^1$ can also be a hydrogen atom but in the case of ketones both groups must contain at least one carbon atom. Hence the simplest ketone contains three carbon atoms.

## Formulae and nomenclature

The formulae and nomenclature follow from the generic names **alkanal** and **alkanone**, as shown in Table 8.2.

**Table 8.2**  Names of aldehydes and ketones

| Aldehydes | |
|---|---|
| **Formula** | **Name** |
| HCHO | methanal |
| $CH_3CHO$ | ethanal |
| $CH_3CH_2CHO$ | propanal |
| $CH_3CH_2CH_2CHO$ | butanal |
| $CH_3CH(CHO)CH_3$ | 2-methylpropanal |

| Ketones | |
|---|---|
| **Formula** | **Name** |
| $CH_3COCH_3$ | propanone |
| $CH_3CH_2COCH_3$ | butanone |
| $CH_3CH_2CH_2COCH_3$ | pentan-2-one |
| $CH_3CH_2COCH_2CH_3$ | pentan-3-one |

Note that there is no alkyl group in methanal but there is a carbon atom in the functional group.

# FUNCTIONAL GROUPS CONTAINING OXYGEN

## Isomerism

Isomerism can occur within each homologous series, as can be seen in the examples butanal is isomeric with 2-methylpropanal, and pentan-2-one is isomeric with pentan-3-one. Within each series, the isomers have the same chemical properties but differ in physical properties such as boiling points etc.

Isomerism can also occur between the two series for molecules with at least three carbon atoms, that is aldehydes are isomeric with ketones. Thus propanal is isomeric with propanone, butanal with butanone (as well as with 2-methylpropanal), etc. In such cases, the isomers will differ in chemical reactions as well as physical properties. A method of distinguishing between the two series is shown in the following section.

## Functional groups and tests

The functional groups are –CHO for aldehydes and $>C=O$ for ketones. Both groups contain a $>C=O$ or carbonyl group.

**Test for the carbonyl group**
This test is for both aldehydes and ketones.
- Reagent. Add an excess of a solution of 2,4-dinitrophenylhydrazine (**Brady's reagent**).
- Result. An orange-yellow precipitate is formed.

**Test for the –CHO group**
There are two possible tests, either of which can be used. These tests are given by aldehydes only.

*Test 1*
- Reagent. Add an ammoniacal solution of silver nitrate and warm. (This solution is made by adding a few drops of dilute sodium hydroxide to a solution of silver nitrate followed by dilute ammonia solution until the brown precipitate dissolves)
- Result. Silver metal is precipitated (often in the form of a mirror on the side of the test tube although it may also appear as a black precipitate).

*Test 2*
- Reagent. Add Fehling's solutions and warm.
- Result. A red precipitate (of copper(I) oxide) is formed.

Both tests for the –CHO group depend on the fact that aldehydes are good reducing agents, while ketones show no reducing properties. The aldehyde is oxidised to carboxylic acid in both cases:

$$-CHO + [O] \rightarrow -CO_2H$$

# Carboxylic acids

## General formula

$$C_nH_{2n+1}CO_2H \quad \text{or} \quad C_mH_{2m}O_2 \text{ (where } m = n+1\text{)}$$

## Formula and nomenclature

Formula and nomenclature follow the generic name **alkanoic acid**, as shown in Table 8.3. Note that the first acid has a value of $n = 0$. This is possible since there is a carbon atom in the $-CO_2H$ group.

**Table 8.3** *The names of some carboxylic acids*

| Formula | Name |
|---|---|
| $HCO_2H$ | methanoic acid |
| $CH_3CO_2H$ | ethanoic acid |
| $CH_3CH_2CO_2H$ | propanoic acid |
| $CH_3CH_2CH_2CO_2H$ | butanoic acid |

## Isomerism

As well as the usual ways of forming isomers, that is, by changing the position of the functional group or branching the carbon chain, carboxylic acids are isomeric with another group of compounds known as esters. This will be discussed later.

## Functional group and its test

The functional group is the carboxyl group.

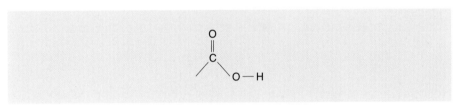

The hydrogen atom bonded to the oxygen reacts differently from those bonded to the carbon atoms. The highly electronegative oxygen atom withdraws electrons from this H, making the loss of this as a proton relatively easy. It is this which makes this series of compounds acidic and this property can be used to test for them.

**Test for the carboxyl group.**
- Reagent. Add to sodium carbonate (or sodium hydrogencarbonate) solution.
- Result. Colourless gas evolved which turns limewater milky.

This test only shows that the compound is acidic and would be given by any acidic compound:

$$CO_3^{2-} + 2H^+ \rightarrow H_2O + CO_2$$
$$\text{or} \qquad HCO_3^- + H^+ \rightarrow H_2O + CO_2$$

It does not in fact prove that a carboxyl group is present. A further test for the —OH group using $PCl_5$ as described earlier is necessary in order to show that a carboxylic acid is present.

## Relationships between alcohols, aldehydes, ketones and carboxylic acids

These homologous series are linked together by a series of redox processes, as shown below.

Primary alcohols, aldehydes and carboxylic acids can be converted.

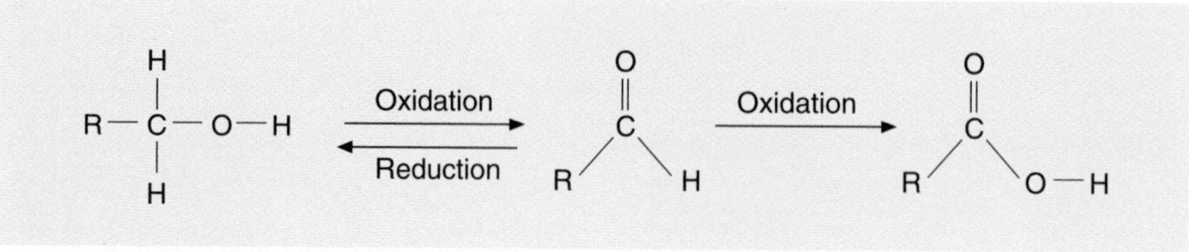

Secondary alcohols and ketones can be interconverted. Ketones resist further oxidation.

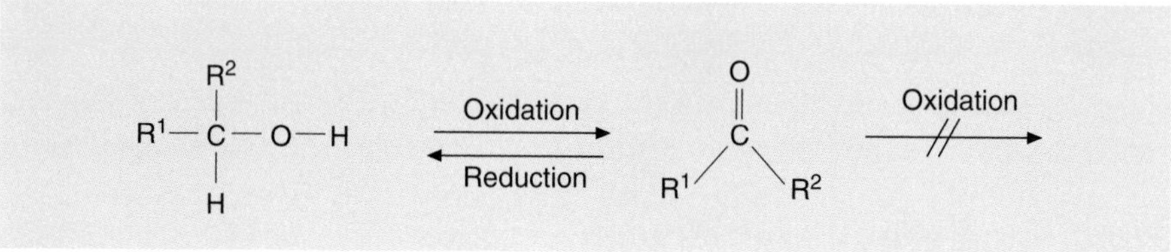

Tertiary alcohols resist oxidation.

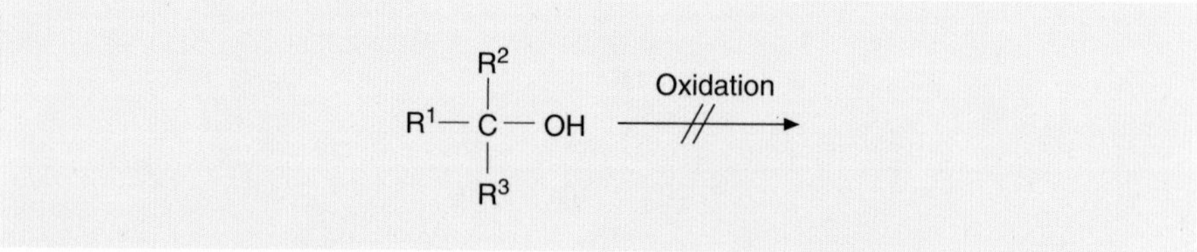

Primary alcohols, secondary alcohols and aldehydes are easily oxidised. Tertiary alcohols and ketones are resistant to oxidation.

### Oxidation reactions

The oxidations shown above can be brought about by a number of oxidising agents. The one which is preferred is potassium dichromate, acidified with dilute sulphuric acid. This gives an easily observed colour change from orange to green.

### *Differentiation between the different types of alcohol*

These oxidation reactions can be used to ditinguish between the different types of alcohol because the aldehydes or ketones produced can be recognised by the tests given earlier. Thus if a liquid

(a) gives 'steamy' fumes on treatment with $PCl_5$

(b) reacts on warming with acidified potassium dichromate to give a product which

(c) gives a positive result with the silver mirror test, then it must be a primary alcohol. Stronger oxidising agents such as potassium manganate(VII) should not be used since they are capable of oxidising some of the simple ketones.

## *Preparation of an aldehyde*

When oxidising a primary alcohol, it would be difficult to stop at the aldehyde stage were it not for the fact that the aldehyde is volatile and can escape before further oxidation occurs. If the reaction is carried out in a distillation apparatus as shown in Figures 8.1(a) and (b), the aldehyde would vaporise and condense so that it can be collected in a separate container. Heating is not necessary in order to perform the oxidation but it is necessary if the aldehyde is the required product. For example:

$$CH_3CH_2OH \; + \; [O] \; \rightarrow \; CH_3CHO + H_2O$$
$$\text{ethanol} \qquad\qquad\qquad \text{ethanal}$$

A primary alcohol can be converted directly to a carboxylic acid in one step by heating under reflux with acidified potassium dichromate.

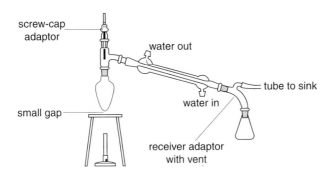

*Fig. 8.1 Distillation apparatus for the preparation of ethanal from ethanol*

## Reduction reactions

### *Aldehydes and ketones*

The reduction of an aldehyde to a primary alcohol or a ketone to a secondary alcohol is relatively easy to achieve using common organic reducing agents such as sodium and ethanol, sodium amalgam and water or metal/acid mixtures. Sodium tetrahydridoborate(III) in water can also be used and this will be dealt with in more detail in module 4. For example:

$$CH_3CHO \; + \; 2[H] \; \rightarrow \; CH_3CH_2OH$$
$$\text{ethanal} \qquad\qquad \text{ethanol}$$

$$CH_3COCH_3 \; + \; 2[H] \; \rightarrow \; CH_3CHOHCH_3$$
$$\text{propanone} \qquad\qquad \text{propan-2-ol}$$

### *Carboxylic acids*

Reduction of carboxylic acids can only be achieved by a reagent known as lithium tetrahydridoaluminate(III). This is similar in its behaviour to sodium tetrahydridoborate(III) but is much more powerful. It also reacts violently with water and so must be used in an inert solvent such as ethoxyethane, carefully dried. Carboxylic acids can only be reduced directly to primary alcohols. For example:

$$CH_3CO_2H \; + \; 4[H] \; \rightarrow \; CH_3CH_2OH \; + \; H_2O$$
$$\text{ethanoic acid} \qquad\qquad \text{ethanol}$$

# FUNCTIONAL GROUPS CONTAINING OXYGEN

## Esters

### General formula

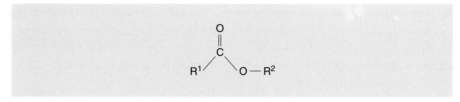

Where $R_1$ and $R_2$ are both alkyl groups $-C_nH_{2n+1}$. The value of $n$ can be 0 for $R_1$ but not for $R_2$ since this would make the compound a carboxylic acid.

### Formulae and nomenclature

Esters are derivatives of carboxylic acids and the stem of the name is **carboxylate**, according to the acid from which they are derived. Hence the stem depends on the nature of $R_1$. For example, ethanoic acid gives esters called ethanoates, methanoic acid gives methanoates, etc. The name of the ester is then completed by a prefix which denotes the name of the alkyl group $R_2$. Some examples are shown in Table 8.4

**Table 8.4** The names of some esters

|   | Formula | Name |
|---|---------|------|
| 1 | $CH_3CO_2CH_3$ | methyl ethanoate |
| 2 | $CH_3CO_2CH_2CH_3$ | ethyl ethanoate |
| 3 | $CH_3CH_2CO_2CH_3$ | methyl propanoate |
| 4 | $HCO_2CH_2CH_3$ | ethyl methanoate |

### Isomerism

There are several examples of isomerism shown within Table 8.4, for example, 1 and 4, or 2 and 3. Apart from this, however, esters are isomeric with carboxylic acids. For example, methyl ethanoate is isomeric with propanoic acid (both are $C_3H_6O_2$) as well as with ethyl methanoate.

Carboxylic acids always have a higher boiling point than the isomeric esters and also respond to the test for an acid (see above). Esters cannot respond to this test, there being no acidic hydrogen atom present.

### Hydrolysis of esters

The term 'hydrolysis' literally means 'reaction brought about by water'. In organic chemistry, however, reaction with water is almost invariably slow even when it is possible. A much faster reaction can be brought about by using a dilute mineral acid or a dilute alkali, but even then heating under reflux is necessary. The reaction of haloalkanes with aqueous sodium hydroxide can therefore be considered to be hydrolysis.

The hydrolysis of esters is best brought about by boiling under reflux with dilute sodium hydroxide. The reaction gives the sodium salt of a carboxylic acid, together with an alcohol. For example:

$$CH_3CO_2CH_2CH_3 + NaOH \rightarrow CH_3CO_2^- Na^+ + CH_3CH_2OH$$

ethyl ethanoate　　　　　　　　sodium ethanoate　　ethanol

## The manufacture of soap

Animal fats are in fact esters made from long chain carboxylic acids (such as stearic acid $C_{17}H_{35}CO_2H$) and the compound glycerol. Glycerol contains three –OH groups and can form a triple ester of the type known as a **glyceride** (as shown in the equation below). Hydrolysis of this type of ester by boiling with sodium hydroxide converts the glyceride into the sodium salt of the stearic acid, which is a **soap**. Indeed the process of hydrolysis of an ester with sodium hydroxide is often referred to as '**saponification**'. The process in general is:

$$
\begin{array}{l}
H_2C - O - \overset{\displaystyle O}{\underset{\displaystyle ||}{C}} - R \\[2mm]
HC - O - \overset{\displaystyle O}{\underset{\displaystyle ||}{C}} - R \quad + 3NaOH \longrightarrow \\[2mm]
H_2C - O - \overset{\displaystyle O}{\underset{\displaystyle ||}{C}} - R \\[2mm]
\qquad\text{a glyceride}
\end{array}
\qquad
\begin{array}{l}
H_2C - O - H \\[2mm]
HC - O - H \qquad + 3RCO_2Na \\[2mm]
H_2C - O - H \\[2mm]
\quad\text{glycerol} \qquad \text{a soap}
\end{array}
$$

The soap is precipitated from solution by addition of sodium chloride. The soap can then be made into bars or into soap powder.

Washing involves the removal of 'dirt', which is basically organic in nature and hence **hydrophobic** or 'water hating'. The problem is that the universal solvent on this planet is water. Soap is effective in removing dirt because it has an ionic end (the carboxylate ion) which is soluble in water (**hydrophilic**) whilst the alkyl chain is soluble in organic matter (for example an oil). Thus water molecules attach themselves to the ionic end of the molecule and the oil molecules attach around the alkyl group. In this way the surface tension between water and the oil is reduced and the miscibility is increased (Figure 8.2).

One disadvantage of soaps is that they they form insoluble calcium salts with the calcium ions in hard water and are therefore wasted to some extent.

## Manufacture of soapless detergents

Soapless detergents work in a similar way to soaps but a sulphonate group ($-SO_2-O^-$) or sulphate group ($-O-SO_2-O^-$) replace the carboxylate group as the hydrophilic component. The calcium salts of these are more soluble in water, and so the problem with hard water is avoided.

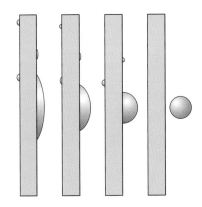

*Fig. 8.2 The effect of detergent solution on fat is to increase the contact angle of the fat with the fibre, so that the fat rolls up into a globule and is detached*

There are several different types of detergent For example, alkylbenzenesulphonates made from non-branched alkenes ($C_{10}-C_{14}$) using HF as a catalyst thus:

$$R{-}CH{=}CH_2 \;+\; C_6H_6 \;\rightarrow\; R{-}CH(CH_3){-}C_6H_5$$
$$\text{benzene} \qquad \text{an alkyl benzene}$$

The alkylbenzene is then heated with concentrated sulphuric acid to form an alkylbenzene sulphonic acid:

$$R{-}CH(CH_3){-}C_6H_5 \;+\; H_2SO_4 \;\rightarrow\; R{-}CH(CH_3){-}C_6H_4{-}SO_3H \;+\; H_2O$$

This is then neutralised to form the sodium salt $R{-}CH(CH_3){-}C_6H_4{-}SO_3{-}Na^+$. This detergent makes up around 10% by weight of most commercial washing powders.

## Synthetic pathways

In Chapters 6–8 several functional groups have been studied to some extent and it is important to realise that the reactions of these functional groups are important for at least two reasons. Firstly, the reactions of the functional groups are always assumed to be the same whether they occur in simple molecules or in more complicated ones. They are also assumed to be the same when there are several functional groups within the same molecule. Secondly, in any reaction of a functional group, a product is formed. Hence the reaction provides a means of making or 'synthesising' the product molecule. The reaction may not work very well in practice but it is at least a possible method of synthesis. The product molecule so formed will be capable of conversion into other molecules, and so on. Hence a series of reactions may be built up to convert one functional group into another. This is called a **synthetic route** or **pathway**.

A summary of synthetic pathways in this module can be found in Figure 8.3.

The conversion of one organic molecule into another may be a simple one-step process or it may involve many steps.

For example, the conversion of ethanol into ethanal is a one-step process achieved by heating the ethanol with acidified potassium dichromate(VI):

$$CH_3CH_2OH \ + \ [O] \ \rightarrow \ CH_3CHO \ + \ H_2O$$

The conversion of 1-bromopropane to propanoic acid, however, involves two steps:

$$\overset{\text{step 1}}{CH_3CH_2CH_2Br \ \rightarrow \ CH_3CH_2CH_2OH} \ \overset{\text{step 2}}{\rightarrow \ CH_3CH_2COOH}$$

Step 1 is achieved by boiling under reflux with aqueous sodium hydroxide. Step 2 is achieved by boiling under reflux with acidified potassium dichromate(VI).

More complex syntheses can be considered after a study of Module 4.

*Fig. 8.3 Synthetic pathways in summary*

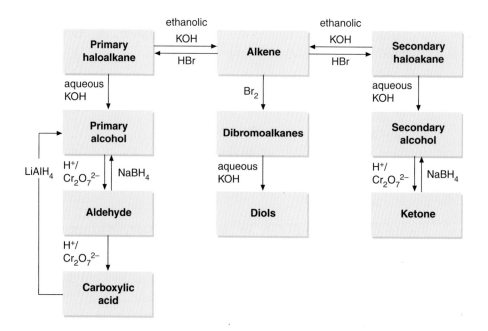

# Questions

1   For each of the homologous series alcohols, aldehydes, ketones and
    carboxylic acids:

    *(a)* write the general formula for the series;

    *(b)* give the formulae and the systematic names of the first three members
    of each series;

    *(c)* give the functional group present in each series;

    *(d)* describe a test for each functional group.


2   *(a)* Write down the structural formulae of all isomers of $C_4H_{10}O$.

    *(b)* Name each isomer.

    *(c)* Identify the different kinds of isomerism occurring and explain how
    each kind of isomerism arises in this particular case.


3   *(a)* Write down the structures of all isomers of $C_2H_4O_2$ and name them.
    Describe a simple chemical test you would perform in order to
    distinguish between these isomers.

    *(b)* Write down the structures of all isomers of $C_4H_8O_2$ which are non-
    acidic.

    *(c)* Write equations for the hydrolysis of each isomer and identify the
    products by name.

    *(d)* Hence suggest how you would distinguish between these isomers
    chemically.


4   Suggest a possible synthesis of:

    *(a)* ethene from ethanal;

    *(b)* ethanal from methyl ethanoate.

# Benzene – the basis of aromatic compounds

The study of organic chemistry is divided into two main sections:
- **aliphatic compounds**; those compounds based on carbon chains,
- **aromatic compounds**; those based on a benzene ring structure.

The compounds studied in previous chapters have all been aliphatic. This chapter will introduce the study of aromatic compounds, which will be continued in Module 4. In order to understand the nature of aromatic compounds it is first necessary to understand the structure of benzene itself.

## Structure of benzene

### The problem

The structure of benzene proved to be a great problem to chemists for many years for the following reasons.
- The molecular formula is $C_6H_6$. This can easily be shown from the percentage composition by mass, determined experimentally .
- The benzene molecule has a regular hexagonal shape and it is planar. This can be shown from a variety of modern techniques such as X-ray diffraction (see Figure 9.4).
- The C–C bond length in benzene is 0.139 nm which is roughly half-way between the C–C bond length in ethane (0.154 nm) and that of C=C in ethene (0.134 nm).
- Benzene undergoes substitution reactions rather than addition.

### The solution

From the knowledge and experience gained in previous chapters, we would expect a molecule such as this to be highly unsaturated and to contain numerous double, or even triple, covalent bonds. One such structure could be

$$CH_2=C=CH–CH=C=CH_2$$

Structures such as this, however, would be expected to undergo numerous addition reactions, but this is not the case for benzene.

### *The Kekulé approach*

The first breakthrough in our understanding of the structure came from a German chemist by the name of Friedrich August Kekulé (1829–96) who was the first to propose a structure in which the carbon atoms form a regular hexagon. His structure was:

*Fig 9.1 Friedrich August Kekulé, who first proposed a structure for benzene*

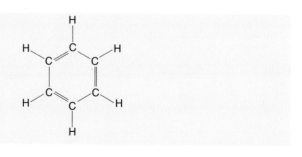

which is represented in the skeletal form as:

This structure looks a little less unsaturated than a chain structure, but it still does not resolve the problems. The shape of the molecule would not be a regular hexagon and the bond lengths would not all be equal. Moreover it would still be expected to undergo addition reactions in the same way as ethene since it appears to have three localised pi bonds. Indeed it might be called 1,3,5-cyclohexatriene.

The problem can be resolved by representing the structure as two forms:

Fig 9.2 A model of the arrangement of the sigma bonds in benzene

where the double-headed arrow is taken to mean that the actual structure lies somewhere in between the two representations. Each C–C bond is therefore neither a single bond nor a double bond, but is of an intermediate type lying between the two forms. Benzene is then described as being a **resonance hybrid** of the two structures. This does NOT mean that the structure oscillates between the two structures but rather that benzene has a single structure in which the C–C bonds are all of the same nature, this being something in between a double bond and a single bond. This concept overcomes the difficulties regarding the shape and the bond lengths of the benzene molecule.

## The molecular orbital approach.

The molecular orbital approach is based on the carbon atoms being $sp^2$ hybridised, with these orbitals being planar and at angles of 120° to one another. Linear overlap between these orbitals could therefore lead to a planar hexagonal structure in which the carbon atoms are all linked by sigma bonds. Similarly, overlap with the 1s orbitals of the six hydrogen atoms would form six more sigma bonds which would also be in the same plane. Each carbon atom still has a p orbital containing a single electron and these protrude above and below the plane of the carbon ring . Each of these p orbitals overlaps laterally with the two p orbitals on either side of it forming a pi molecular orbital. This molecular orbital is different from those seen in alkenes, however, since it is not formed by just two atomic orbitals but by six. The molecular orbital is therefore not in one position between two particular C atoms but is spread all round the carbon ring and is said to be **delocalised**. A model of benzene based on this approach to the structure is shown in Figure 9.3.

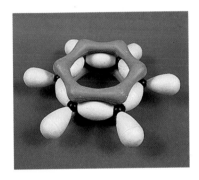

Fig 9.3 A model of the structure of the benzene ring showing the delocalised π bond

An electron density contour map, obtained by X-ray diffraction, is shown in Figure 9.4 and this clearly demonstrates that the high electron density is spread equally around the whole ring and not concentrated in particular areas.

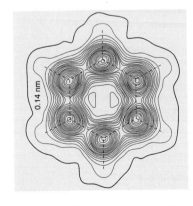

Fig. 9.4 An electronic density contour map of benzene obtained by X-ray diffraction

Both the Kekulé structure and the delocalised structure have advantages and disadvantages if organic chemistry is studied to a higher level. We cannot be sure which of these two approaches to the structure of benzene is correct. Both approaches lead to acceptable representations for the skeletal structure of benzene, but what must not be used is the cyclohexane structure ⬡.

## Thermochemical evidence for the structure of benzene.
The enthalpy changes associated with certain reactions provides some powerful evidence in support of the structure proposed for benzene above.

### Enthalpies of hydrogenation
The molecule cyclohexene can be hydrogenated to form cyclohexane:

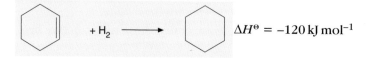

Benzene is also capable of hydrogenation to form cyclohexane. If the structure of benzene was 1,3,5-cyclohexatriene (a Kekulé-type structure), the reaction and the expected enthalpy change would be:

When measured experimentally the actual enthalpy change for the hydrogenation of benzene is $-208\,kJ\,mol^{-1}$, which is considerably lower than the value predicted from a structure containing three localised double bonds. If this information is put on an energy diagram as in Figure 9.5, it can be seen that the actual structure of benzene is more stable than a structure based on localised double bonds by $152\,kJ\,mol^{-1}$.

### Enthalpies of atomisation
The enthalpy of atomisation of 1,3,5-cyclohexatriene (a Kekulé structure) is represented by the reaction:

$\Delta H$ for this reaction can be calculated from average bond enthalpies ($E$) as shown below:

$$\Delta H = [3 \times E(C=C)] + [3 \times E(C-C)] + [6 \times E(C-H)]$$

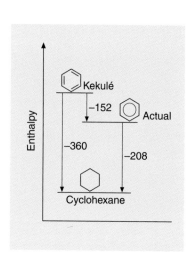

*Fig. 9.5 Enthalpy diagram, showing enthalpies of hydrogenation for benzene*

Inserting values of average bond enthalpies from data books gives:

$$\Delta H = (3 \times 612) + (3 \times 348) + (6 \times 412)$$
$$= 1836 + 1044 + 2472$$
$$= +5352 \, \text{kJ mol}^{-1}$$

The enthalpy of atomisation of benzene based on the actual structure can be calculated from an energy cycle, by using the enthalpies of formation of benzene, H(g) and C(g).

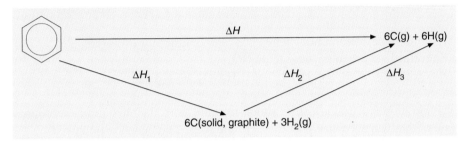

Application of Hess's law gives:

$$\Delta H = \Delta H_1 + \Delta H_2 + \Delta H_3$$

Inserting values from data books gives:
$\Delta H_1 = -49 \, \text{kJ mol}^{-1}$; $\Delta H_2 = +4302 \, \text{kJ mol}^{-1}$; $\Delta H_3 = +1308 \, \text{kJ mol}^{-1}$.

$$\Delta H = (-49) + (+4302) + (+1308) = +5561 \, \text{kJ mol}^{-1}.$$

Comparing this value with the value obtained for the theoretical Kekulé structure on the energy diagram in Figure 9.6, again shows the actual structure of benzene to be more stable than a single Kekulé structure, this time by 209 kJ mol$^{-1}$. This compares well with the value of 152 kJ mol$^{-1}$ obtained earlier, considering that average bond enthalpies were used.

Either method shows quite clearly that however the structure of benzene is represented, it is energetically much more stable than a structure containing localised pi bonds.

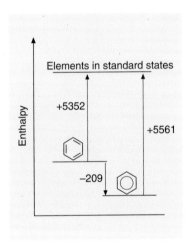

*Fig. 9.6 Enthalpy diagram, showing enthalpies of atomisation for benzene*

### Reason for substitution reactions rather than addition

The one remaining dilemma in the problems associated with the structure of benzene can now be resolved. The benzene ring is thermodynamically very stable as we have shown in the previous section and this stability is associated with the delocalisation of the pi molecular orbital. Addition reactions to benzene would result in disruption of this delocalisation and so reduce the stability of the benzene ring. Substitution reactions can occur without any such disruption and so the stability of the benzene ring is maintained. One such substitution reaction will be considered in the next section. Others will be dealt with in Module 4.

## Nitration of benzene

### What is nitration?

Nitration is the term used to describe the introduction of a $-NO_2$ group (a nitro group) into a molecule. This type of reaction is possible in both aliphatic and aromatic compounds, but it is much easier to achieve, and is probably of greater importance, in aromatic compounds.

### The nitration of benzene

- Reagents.    A mixture of concentrated nitric acid and concentrated sulphuric acid.
- Conditions.  Warm under reflux at a temperature not exceeding 55°C.
- Product.     Nitrobenzene ($C_6H_5NO_2$)
- Equation.    $C_6H_6 + HNO_3 \rightarrow C_6H_5NO_2 + H_2O$

The nitration reaction is a substitution reaction, with the nitro group replacing one of the hydrogen atoms of the benzene ring. The benzene ring itself remains intact and is as fully delocalised after the reaction as it was before reaction. Hence its thermodynamic stability has not been reduced.

### The practical technique in outline

**This is only an outline preparation and must not be used as laboratory instructions. Before attempting the preparation in the laboratory consult a manual of practical chemistry and your teacher.**

The concentrated acids must be mixed together before being added to the benzene. This may result in a violent reaction and so must be done with great care. The concentrated sulphuric acid is added in small portions to the concentrated nitric acid (not the other way round) with constant stirring and cooling. When the mixing is complete, the mixture is cooled and then added carefully in small portions to the benzene, shaking well after each addition and cooling if necessary. Finally the mixture is warmed in a reflux apparatus, using a bath of warm water for heating the flask.

## Mechanism for the nitration of benzene

The electron-rich, delocalised pi bond system in benzene is susceptible to attack by electrophiles, that is, species which are capable of accepting a pair of electrons. As we have already seen, the reaction occurs by substitution, so this type of mechanism is referred to as **heterolytic electrophilic substitution**.

This mechanism occurs in two stages.

### Stage 1: generation of the electrophile

When the acids are mixed, a reaction occurs forming the electrophile $NO_2^+$ which is known as the nitronium ion:

$$HNO_3 + H_2SO_4 \rightarrow H_2NO_3^+ + HSO_4^-$$

$$H_2NO_3^+ + H_2SO_4 \rightarrow H_3O^+ + HSO_4^- + NO_2^+$$

Sulphuric acid, being a stronger acid, donates a proton to the nitric acid (which is therefore acting as a base), forming its conjugate acid $H_2NO_3^+$. This then loses a molecule of water to form the nitronium ion.

If preferred, a single equation can be written in place of the two above:

$$HNO_3 + 2H_2SO_4 \rightarrow H_3O^+ + 2HSO_4^- + NO_2^+$$

### Stage 2: electrophilic attack on the benzene ring

The benzene ring donates an electron pair from the delocalised pi bond system to form a covalent bond with the electrophile $NO_2^+$. The ring is now only partially delocalised and is positively charged, and so it is less stable. Loss of a proton from this unstable intermediate feeds two electrons back to the benzene ring and full delocalisation is restored. The net effect is that an $H^+$ ion is displaced from the benzene ring and replaced by an $NO_2^+$ ion. The whole process is represented as:

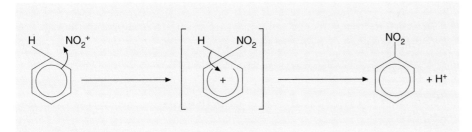

# Questions

1   (a) The enthalpy change for the reaction

$$CH_2{=}CH_2 + H_2 \rightarrow CH_3CH_3$$

is $-120\,kJ\,mol^{-1}$, whereas that for the reduction of benzene, $C_6H_6$, to cyclohexane, $C_6H_{12}$, is $-208\,kJ\,mol^{-1}$. What may be deduced from the fact that this value is not three times the first one?

   (b)   (i) State the conditions under which benzene may be nitrated to form  mononitrobenzene

   (ii) Both of the reagents that are used to nitrate benzene are usually regarded as acids. However, in this instance, one of them behaves as a base. Show how this is so.

   (iii) Give the mechanism for the nitration of benzene.

   (iv) Explain why benzene tends to undergo substitution rather than addition reactions.

*(ULEAC 1995)*

2   The relative rates of hydrolysis of some bromoalkanes by hydroxide ions are shown below.

| Bromoalkane | Relative rate of hydrolysis |
|---|---|
| $CH_3-Br$ | 1.8 |
| $CH_3CH_2-Br$ | 1.1 |
| $(CH_3)_2CH-Br$ | 1.0 |
| $(CH_3)_3C-Br$ | 2.7 |

*(a)*  (i)  Show the $S_N1$ mechanism for the hydrolysis of $(CH_3)_3C$-Br.

(ii)  How do you account for the relative rates of hydrolysis of $(CH_3)_2CH-Br$ and $(CH_3)_3C-Br$, which proceed by the same mechanism?

*(b)*  The chiral bromoalkane $C_2H_5CH(CH_3)Br$ gives rise to a racemic mixture on hydrolysis by hydroxide ion.  What can you deduce about the mechanism of hydrolysis of this chiral bromoalkane?

*(ULEAC 1993 (part)).*

3   *(a)*  Give the reagents and conditions necessary to bring about the following changes for 2-bromobutane.

(i)  $CH_3CH_2CHBrCH_3 \rightarrow CH_3CH_2CH=CH_2$
     A                         B

(ii)  $CH_3CH_2CHBrCH_3 \rightarrow CH_3CH_2CHOHCH_3$
      A                         C

*(b)*  In part (a)(i) an alkene other than B is formed.

(i)  Identify this alkene.

(ii)  State one type of isomerism shown by this compound which is not shown by B.  Briefly explain how this occurs.

*(c)*  A can be converted into an acid E in a two stage process via compound D.

$CH_3CH_2CHBrCH_3 \rightarrow CH_3CH_2CHCNCH_3 \rightarrow CH_3CH_2CH(CO_2H)CH_3$
          A                           D                              E

(i)  Name compound E

(ii)  What type of isomerism is shown by both A and E ?

(iii)  Give the reagents and conditions for the conversion of A to D.

(iv)  Give the mechanism for the conversion of A to D

*(ULEAC 1995(amended))*

# Numerical answers

## Chapter 1

1  (b)  $-393.5 \, \text{kJ mol}^{-1}$; $-285.9 \, \text{kJ mol}^{-1}$

2  (b)  $-416.6 \, \text{kJ mol}^{-1}$

3  $-126.4 \, \text{kJ mol}^{-1}$

4  (b)  $+0.4 \, \text{kJ mol}^{-1}$

5  (b)  ethane: $-1558.9 \, \text{kJ mol}^{-1}$;
      ethene: $-1409 \, \text{kJ mol}^{-1}$;
      hydrogen: $-285.5 \, \text{kJ mol}^{-1}$

   (c)  $-135.6 \, \text{kJ mol}^{-1}$

6  (a)  $-45.8 \, \text{kJ mol}^{-1}$

   (b)  $+90 \, \text{kJ mol}^{-1}$

7  (a)  $-74 \, \text{kJ mol}^{-1}$

   (b)  $-109 \, \text{kJ mol}^{-1}$

   (c)  $-606 \, \text{kJ mol}^{-1}$

8  diborane: $-3.67 \times 10^6 \, \text{kJ}$; benzene: $-4.06 \times 10^6 \, \text{kJ}$

9  $+333.7 \, \text{kJ mol}^{-1}$

10  $+31 \, \text{kJ mol}^{-1}$

11  $+390.8 \, \text{kJ mol}^{-1}$

## Chapter 2

1  (a)  (i) $x$;   (ii) $x - y$ (not $y - x$)

   (c)  $x - y$

3  (b)  Rate $= k[\text{OH}^-][\text{CH}_3\text{CSNH}_2]$

   (c)  Rate doubles

4  (e)  2; Rate $= k[\text{HCl}]^2$

5  (b)  bromate $= 1$; bromide $= 1$; $\text{H}^+ = 2$

   (c)  Rate $= k[\text{BrO}_3^-][\text{Br}^-][\text{H}^+]^2$;
        Units of $k = \text{mol}^{-3} \text{dm}^9 \text{s}^{-1}$

## Chapter 3

1  (a)  Right

   (b)  Stays same

   (c)  Right

   (d)  Left

4  $[\text{H}_2] = [\text{I}_2] = 0.6$; $[\text{HI}] = 4.8 \, \text{mol dm}^{-3}$

5  $[\text{C}_2\text{H}_5\text{OH}] = 0.333$;
   $[\text{CH}_3\text{CO}_2\text{C}_2\text{H}_5] = [\text{H}_2\text{O}] = 0.667$; $K_c = 4$

6  (c)  $550\,^\circ\text{C}$; 200 atm

7  (b)  54.7 no units

## Chapter 4

2  (a)  $[\text{H}^+] = 1.34 \times 10^{-3} \, \text{mol dm}^{-3}$

   (b)  pH $= 2.87$

3  (a)  3

   (b)  2.7

   (c)  11.3

   (d)  2.6

4  (a)  (i) $0.001 \, \text{mol dm}^{-3}$;   (ii) $0.0063 \, \text{mol dm}^{-3}$

   (b)  $[\text{H}^+] = 10^{-11} \, \text{mol dm}^{-3}$;
        $[\text{NaOH}] = 0.001 \, \text{mol dm}^{-3}$

5  (b)  pH $= 4.6$

6  (a)  [Acid] : [Salt] $= 1 : 1.58$

   (b)  $[\text{CH}_3\text{CO}_2\text{H}] : [\text{CH}_3\text{CO}_2^-] = 1 : 2$
        e.g. $[\text{CH}_3\text{CO}_2\text{H}] = 0.1 \, \text{mol dm}^{-3}$;
        $[\text{CH}_3\text{CO}_2^-] = 0.2 \, \text{mol dm}^{-3}$

## Chapter 5

5  (b)  $\text{C}_2\text{H}_6\text{O}$; $\text{CH}_3\text{CH}_2\text{OH}$ and $\text{CH}_3\text{OCH}_3$

# Examination questions

**1** (*a*) (i) Ammonia is a base. What is meant by the term *base*?

(ii) What feature of the ammonia molecule enables it to react as a base?

(iii) Given that $Kb = \dfrac{[OH^-][NH_4^+]}{[NH_3]}$ calculate the pH of a 0.100 mol dm$^{-3}$ solution of ammonia.

The values of $K_b$ and $K_w$ are $1.80 \times 10^{-5}$ mol dm$^{-3}$ and $1.00 \times 10^{-14}$ mol$^2$ dm$^{-6}$ respectively. **(5)**

(*b*) (i) Sketch on the axes below the pH curve for the titration of 25 cm$^3$ of 0.1 mol dm$^{-3}$ HCl with 0.1 mol dm$^{-3}$ ammonia.

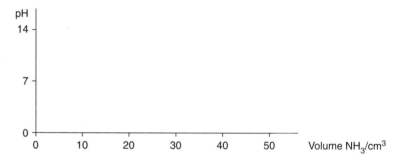

(ii) What indicator would you use for this titration? **(3)**

(*c*) There is no suitable indicator for the titration of ethanoic acid with ammonia. Why is this? **(2)**

(*d*) Suggest the conditions under which ammonia could be used to convert $CH_3CHClCO_2H$ to $CH_3CH(NH_2)CO_2H$. **(2)**

(*e*) Explain how 2-aminopropanoic acid, $CH_3CH(NH_2)CO_2H$, can act as a buffer in solution. **(3)**

**Total 15 marks**
*(ULEAC GCE Chemistry (9081/8081), June 1996)*

**2** Compound **X** has the molecular formula $C_5H_8O$.

(*a*) Calculate the percentage composition of **X**. **(2)**

(*b*) Give the reagents you would use to show the presence of each of the following groups which are present in **X**. State what you would observe as the result of each test.

(i) $\diagdown C = C \diagup$          (ii) $\diagdown C = O$     **(4)**

(c) Given that **X** also forms a silver mirror when warmed with ammoniacal silver nitrate solution and shows no geometrical isomerism, write TWO structures for **X**, neither of which is chiral. **(2)**

(d) Select ONE of your structures for **X**. This reacts with HBr to form two products, one of which is chiral.

  (i) Give the structures of your two products, indicating which is chiral.

  (ii) Write the mechanism for the formation of either product.

  (iii) Indicate which of your two products is likely to be in the greater yield. Give a reason for your choice. **(7)**

**Total 15 marks**

*(ULEAC GCE Chemistry (9081/8081), June 1996)*

**3**  This question is about chlorofluorocarbons (CFCs) and hydrofluoroalkanes (HFAs).

(a) Give TWO equations which show the mechanism by which the reaction of chlorine with methane in dull sunlight is propagated after the initial formation of chlorine free radicals:

$$Cl_2 \rightarrow Cl^\bullet + Cl^\bullet$$  **(2)**

(b) Chlorofluorocarbons have been used in refrigeration and in aerosols. Suggest ONE property which has made such compounds useful. **(1)**

(c) Give the structural formula, showing all covalent bonds, for the CFC 1,1,2-trichloro-1,2,2-trifluoroethane. **(1)**

(d) CFCs lead to the destruction of ozone in the stratosphere; these compounds form chlorine monoxide and chlorine free radicals which convert ozone to oxygen. What is the environmental significance of destroying the ozone layer? **(1)**

(e) Hydrofluoroalkanes are being developed to replace CFCs. One such compound is 1,1,1,2-tetrafluoroethane. Why should this compound not cause the destruction of ozone? **(1)**

(f) 1,1,1,2-tetrafluoroethane can be made by a two-stage synthesis. Write equations for the following reactions.

  (i) Tetrachloroethene is reacted with chlorine and hydrogen fluoride to produce 1,1-dichlorotetrafluoroethane and hydrogen chloride.

  (ii) The 1,1-dichlorotetrafluoroethane is reduced by hydrogen gas to 1,1,1,2-tetrafluoroethane and hydrogen chloride. **(2)**

**Total 8 marks**

*(ULEAC GCE Chemistry (Advanced Supplementary), June 1995)*

# EXAMINATION QUESTIONS

**4** (*a*) (i) Define pH.

    (ii) Give an expression for the dissociation constant, $K_a$, for an acid HX. **(2)**

  (*b*) (i) When $50.0\,cm^3$ of $1.00\,mol\,dm^{-3}$ hydrochloric acid was mixed with $50.0\,cm^3$ of $1.00\,mol\,dm^{-3}$ sodium hydroxide, the temperature rose by $6.82\,°C$. Calculate the enthalpy of neutralisation of hydrochloric acid by sodium hydroxide.

      **Data:** All the above solutions may be assumed to have a specific heat capacity of $4.2\,JK^{-1}cm^{-3}$.

      Heat gain or loss by solution

      = volume × specific heat capacity × temperature change.

    (ii) The enthalpy of neutralisation for ethanoic acid and sodium hydroxide is $-55.2\,kJ\,mol^{-1}$. Comment on this value in relation to that found in (i). **(5)**

  (*c*) Many acid-base indicators are weak acids and may be written as HInd. The dissociation of the indicator bromophenol blue may be represented as follows:

$$HInd(aq) \rightleftharpoons H^+(aq) + Ind^-(aq)$$
    yellow                blue

    The acid (or indicator) dissociation constant is denoted by $K_{Ind}$.

    (i) Given that $K_{Ind}$ is $1.00 \times 10^{-4}\,mol\,dm^{-3}$ for bromophenol blue, find the pH for which the indicator would be green, i.e. when there are equal concentrations of the yellow and blue species present.

    (ii) The range of pH over which the indicator changes colour is given as 2.8–4.6. Calculate the ratio $\dfrac{[HInd]}{[Ind^-]}$ at pH 2.8.

    (iii) What will be the indicator colour at pH 2.8? Give a reason for your answer. **(5)**

  (*d*) State how and under what conditions

    (i) ethanoic acid reacts with ethanol;

    (ii) chloroethanoic acid reacts with aqueous potassium hydroxide. **(6)**

**Total 18 marks**
*(ULEAC GCE Chemistry (Advance Supplementary), June 1995)*

**5** (*a*) (i) Define pH

    (ii) Define $k_w$, the ionic product of water. **(2)**

(b) Calculate the pH of the following solutions. (The ionic product of water $K_w$, may be taken as $1.00 \times 10^{-14} \, mol^2 \, dm^{-6}$.)

    (i) A solution of sulphuric acid having a concentration of $0.100 \, mol \, dm^{-3}$.

    (ii) A solution of sodium hydroxide having a concentration of $0.0500 \, mol \, dm^{-3}$. **(3)**

(c)   (i) What is the principal property of a buffer solution?

    (ii) The dissociation constant for ethanoic acid is $1.80 \times 10^{-5} \, mol \, dm^{-3}$. Calculate the pH of a buffer solution which has a concentration of $0.0150 \, mol \, dm^{-3}$ with respect to ethanoic acid and $0.0550 \, mol \, dm^{-3}$ with respect to sodium ethanoate. **(5)**

(d) Phosphoric(V)acid, $H_3PO_4$, is a tribasic acid.

Write the formulae of the potassium salts of this acid. **(1)**

(e) $25.0 \, cm^3$ of a solution of phosphoric(V) acid of concentration $0.0590 \, mol \, dm^{-3}$ required $28.1 \, cm^3$ of $0.105 \, mol \, dm^{-3}$ sodium hydroxide for reaction in a titration, using phenolphthalein indicator.

    (i) Calculate the number of moles of phosphoric(V) acid in $25.0 \, cm^3$.

    (ii) Calculate the number of moles of sodium hydroxide in $28.1 \, cm^3$.

    (iii) Write a balanced equation for the reaction of phosphoric(V) acid with sodium hydroxide under these conditions. **(4)**

**Total 15 marks**

*(ULEAC GCE Chemistry (Advanced/Supplementary Module Test 2),*
*January 1996)*

**6** (a) Sketch on axes labelled 'reaction co-ordinate' (x-axis) and 'energy' (y-axis) the reaction profiles (enthalpy level diagrams) for

    (i) a single stage exothermic reaction,

    (ii) a two-stage reaction in which an intermediate compound is formed, the first stage being endothermic and the overall reaction being exothermic. **(4)**

(b) Bromide, bromate(V) and hydrogen ions react according to the following equation:

$$6H^+(aq) + 5Br^-(aq) + BrO_3^-(aq) \rightarrow 3Br_2(aq) + 3H_2O(l)$$

The reaction may be carried out in the presence of small amounts of phenol and methyl orange. When sufficient bromine is formed to use up all the phenol, the methyl orange is decolorised by the bromine. The time taken for this decolorisation may be used to calculate the initial rate of reaction. Results obtained in such an experiment were:

| Experiment number | Initial concentration of $Br^-$/ mol dm$^{-3}$ | Initial concentration of $H^+$/ mol dm$^{-3}$ | Initial concentration of $BrO_3^-$/ mol dm$^{-3}$ | Relative initial rate |
|---|---|---|---|---|
| 1 | 0.00278 | 0.0333 | 0.00139 | 4 |
| 2 | 0.00139 | 0.0333 | 0.00139 | 2 |
| 3 | 0.00278 | 0.0333 | 0.00069 | 2 |
| 4 | 0.00278 | 0.0167 | 0.00139 | 1 |

(i) Find the order of the reaction with respect to each of the following, giving your reasoning: $Br^-$ $H^+$ $BrO_3^-$

(ii) Write a rate equation for the reaction.

(iii) What is the overall order of the reaction? **(6)**

(c) If the volume of the mixture is 40 cm$^3$, calculate for Experiment 1:

(i) the number of moles of $BrO_3^-$ present in the original mixture;

(ii) the number of moles of bromine liberated if the reaction goes to completion. **(3)**

(d) (i) Bromine and phenol react as follows:

$$C_6H_5OH(aq) + 3Br_2(aq) \rightarrow Br_3C_6H_2OH(aq) + 3H^+(aq) + 3Br^-(aq)$$

The number of moles of phenol present in the original mixture of every experiment is $1.00 \times 10^{-6}$. How many moles of bromine have been produced at the colour change?

(ii) Express this number of moles of bromine as a percentage of that formed in (c)(ii). **(2)**

**Total 15**

*(ULEAC GCE Chemistry (Advanced/Supplementary Module Test 2), January 1996*

7 (a) Define the term **standard enthalpy of combustion**. **(3)**

(b) Calculate the enthalpy change for the reduction of propanoic acid to propanal:

$$CH_3CH_2COOH + H_2 \rightarrow CH_3CH_2CHO + H_2O$$

given the following enthalpies of combustion /kJ mol$^{-1}$:

propanoic acid – 1527; hydrogen – 286; propanal –1821. **(3)**

(c) Propanoic acid is a *weak acid*; explain the term **weak**. **(1)**

(d) (i) Give an equation for the dissociation of propanoic acid and hence an expression for its dissociation constant, $K_a$.

(ii) At 25 °C $K_a$ for propanoic acid is $1.30 \times 10^{-5}\,mol\,dm^{-3}$. Find the pH of a solution of propanoic acid of concentration $0.0100\,mol\,dm^{-3}$. State any assumptions you make.

(iii) Increasing the temperature of the propanoic acid solution causes the pH to decrease. What does this tell you about the enthalpy of dissociation? Justify your answer. **(9)**

**Total 16 marks**

*(ULEAC GCE Chemistry (Advanced/Supplementary Module Test 2), June 1995)*

**8** Ammonia is manufactured from hydrogen and nitrogen in the Haber process:

$$N_2(g) + 3H_2(g) \rightleftharpoons 2NH_3(g) \quad \Delta H = -92\,kJ\,mol^{-1}$$

(*a*) What is meant by the term **dynamic equilibrium**? **(2)**

(*b*) State the conditions employed industrially in the manufacture of ammonia, and justify them on physico-chemical grounds. **(5)**

(*c*) What effect does a catalyst have on the rate of achievement of the equilibrium and the composition of the equilibrium mixture? **(2)**

(*d*) A significant proportion of the ammonia made is oxidised to nitrogen monoxide and steam, using oxygen and a platinum/rhodium catalyst.

(i) Give the equation for the reaction.

(ii) Use the enthalpies of formation, $\Delta H_f$, given below to evaluate $\Delta H$ for the oxidation of ammonia.

|           | $\Delta H_f/kJ\,mol^{-1}$ |
|-----------|------------------|
| $NO(g)$   | +90              |
| $H_2O(g)$ | −242             |
| $NH_3(g)$ | −46              |

(iii) Ammonia does not spontaneously catch fire in air or oxygen; use this fact and your result from part (ii) to explain the difference between **kinetic** and **thermodynamic** stability. **(7)**

**Total 16 marks**

*(ULEAC GCE Chemistry (Advanced/Supplementary Module Test 2), June 1995)*

# Index